Perspectives on Kentucky's Past

ARCHITECTURE, ARCHAEOLOGY, AND LANDSCAPE

Julie Riesenweber
General Editor

Rock Fences of the Bluegrass

Carolyn Murray-Wooley
and
Karl Raitz

Color Photographs by Ron Garrison

THE UNIVERSITY PRESS OF KENTUCKY

Frontispiece: Photograph by Ron Garrison

Scholarly publisher for the Commonwealth,
serving Bellarmine College, Berea College, Centre
College of Kentucky, Eastern Kentucky University,
The Filson Club, Georgetown College, Kentucky
Historical Society, Kentucky State University,
Morehead State University, Murray State University,
Northern Kentucky University, Transylvania University,
University of Kentucky, University of Louisville,
and Western Kentucky University.

Editorial and Sales Offices: Lexington, Kentucky 40508-4008

Library of Congress Cataloging-in-Publication Data

Wooley, Carolyn Murray.
Rock fences of the bluegrass / Carolyn Murray-Wooley and Karl Raitz.
p. cm. — (Perspective on Kentucky's past)
Includes bibliographical references and index.
ISBN 0-8131-1762-3
1. Stone walls—Kentucky. 2. Limestone—Kentucky. I. Raitz, Karl B. II. Title. III. Series.
TH2249.W66 1992
693′.1—dc20 91-22584

This book is printed on recycled acid-free paper meeting
the requirements of the American National Standard
for Permanence of Paper for Printed Library Materials.
♾

Contents

Sponsor's Foreword

Kentucky's thousands of cultural resources form a tangible record of twelve thousand years of history and prehistory. They include archaeological sites such as native American villages and burial mounds, the historic remains of fortifications of our first European settlers, and Civil War earthworks and battlefields. Above ground are structures ranging from individual houses to entire streetscapes of Victorian commercial buildings. These resources combine to form a past and present environment—a cultural landscape—worthy of preservation.

Preservationists have always made decisions about which cultural resources should remain for future generations, but these decisions are becoming even more difficult. No longer is preservation a simple matter of saving old buildings from the wrecking ball or restoring them to their original appearance. Preservationists today must not only consider a more comprehensive and diverse array of properties, but also attempt to unravel the complex relationships among them. One way the profession responds to challenges posed by the rapidly changing times is to seek better understanding of the cultural resources with which it is concerned.

The State Historic Preservation Office, the Kentucky Heritage Council, encourages study of the commonwealth's architecture, archaeology, and landscape. As a growing number of constituents demand that decisions be weighed in light of many special interests, preservation increasingly becomes a public endeavor. The profession must also find ways of communicating information gained through research. This volume inaugurates the latest of the Kentucky Heritage Council's ongoing educational efforts—the Perspectives on Kentucky's Past series. By sharing the insights of current scholarship, the council seeks ultimately to develop the broad support necessary to preserve buildings, sites, and structures important in the state's past.

Rock Fences of the Bluegrass is an ideal book with which to open the series since it is the first scholarly treatment of a common, but little understood, element of our cultural landscape. Karl Raitz and Carolyn Murray-Wooley here combine the perspectives of geography and architectural history to document changes in the structure, function, and meaning of these fences from their initial popularity in the mid-nineteenth century to the present. Their research utilizes written records, oral history, and close examination of many fences. In charting the transmission of fence-building skills from one group of craftsmen to another, the authors make strong connections between the past and present and the Old World and the New. They also link the fences, built of the limestone abundant in central Kentucky's Bluegrass region, with the land, making a strong argument that these symbolize important themes in the state's history and should be preserved.

David L. Morgan
Executive Director
Kentucky Heritage Council

Acknowledgments

In 1988, we began field work and archival research on the geography and history of Kentucky's rock fences. The importance of the fences as landscape artifacts and benchmarks of cultural history became apparent as we documented rock fence construction technology and fence-building history. The scarcity of readily available information about the fences and concern for their preservation prompted this book.

Many people aided in this work. State Historic Preservation officer David L. Morgan and survey coordinator Robert M. Polsgrove, both of the Kentucky Heritage Council, gave initial direction for the project by providing a grant to develop a methodology for nominating rock fences to the National Register of Historic Places.

Dorn VanDommelen served as research assistant on the project during the summer of 1988 and contributed aerial photo interpretation and map drafting. Christopher Williams and Bill Kephart of Shakertown at Pleasant Hill provided useful citations from the Shaker records at the Filson Club Library in Louisville. Ann Medaris wrote to tell us of rock fence information in the Fayette County Order Books. Staff at the M.I. King Library Special Collections at the University of Kentucky aided our research, especially Betty J. Gooch and James D. Birchfield. County extension agents took time to indicate on highway maps places of notable fence concentrations in their counties; those agents include William Brinkley, George Dennis Cantrill, Maner Ferguson, Bernie C. Milam, David Sparrow, Roger Sparrow, Kim Strohmeier, and John K. Wills. Stanley Kelly, restoration craftsman, generously allowed us to read and quote his manuscript, "What Yesterday Left Us," describing his discoveries about early Kentucky buildings during the process of restoration.

We conducted interviews and correspondence with many people who provided useful information. They include Alice Algood, Linda Anderson, Martha B. Bogges, Elva Bradley, Mary C. Breeding, Mrs. Robert Brewer, Joe Brown, Judy Brumfield, Wayne Carmickle, Samuel Cassidy, J.E. Caton, Mank Chatfield, William H. Chatfield, Bill Clark, Norton Clay, Edward Conner, Rebecca Conover, Evelyn Taylor Cook, Alice and J.R. Cox, Dan Cox, John Creech, Jerry Cutsinger, Marjorie Davenport, Garland Dever, Hazel Etherington, Linda Fannin, Andrew Finn, Mrs. Ritchie Flynn, Ted Garrison, David Gear, Ben Giles, Dan Goodman, Howard Gregory, Ann Grundy, John H. Guy III, Eli C. Hall, Neal O. Hammon, Paul Harp, Mexico Hayden, Jay Hensley, Jack Higgins, Benny Higgs, Charles P. Hockensmith, Paul Holleran, James Huff, Mrs. Nick Huff, Howard Humphrey, Mike Huskisson, Doris Isham, Mack Jackson, Frances Keightley, Clay Lancaster, Beverly J. Lasher, Charles R. Lee, George Letton, Mary F. May, Joan Mayer, Bernice McClanahan, Julie Schmitt Metzger, David Miles, Miles Miller, Don Mogge, Tom Moore, Rena Niles, John Penn, Stephen Price, Tim Prosser, Trudie Reed, Virginia Reed, Florence Rhinehart, James T. Richardson, James S. Rush, Elizabeth Sea, Forrest Sea, Kay Shumate, Marvin T. Smith, William Bradley Smith, William W. Smith, John Soper, Frances Thornton Spengler, Thelma Standiford, Sarah House Tate, Ed Taylor, Richard Taylor, James C. Thomas, James Ed Thornton, Retha Traynor, Jim Tuttle, John Venable, Larry Waugh, Ron Wells, Howard Wiles, George F. Willmott, Jr., Mary Allan Wilson, Malcolm Witt, Pat Wolcott, Robert Wolfe, and Charles Worford.

Several people kindly granted us extended interviews, allowed access to family records, research and manuscripts, or accompanied us on tours of their property for fence measurement and photography. To these we offer a special thank you: Ann Bolton Bevins, Joe Brown, Clemens Caldwell, Thomas D. Clark, Berle Clay, Paul J. Gormley, Paul Harp, Susan Hinkle, J.R. Miller, Mary Stuart Newman, Tom Soper, and Bobby Waugh.

Information on rock fence building and preservation techniques in Ireland and the United Kingdom came from J.R.R. Adams of the Ulster Folk and Transport Museum, Richard Conniff, Patricia Donlon of the National Library of Ireland, Elsie and Ted Ellwood, Dave Goulder, P.B. Griffiths of the National Trust of Great Britain, Mosman Haddow, J. Hill, Simon Hodgson of the British Trust for Conservation Volunteers, A. Jones, Patricia Lysaght of the University College, Dublin, Douglas Mathieson of the National Library of Scotland, Caomhin O Danachair, Philip Robinson of the Ulster Folk and Transport Museum, and Roy Stanley of Trinity College, Ireland. Jacqui Simkins of the Dry Stone Walling Association of Great Britain kindly canvased the membership for details comparing British Isles and Kentucky fence construction. Richard Tufnell, master dyker, corresponded with us at length and generously proofread our glossary and corrected errors in interpretation.

Ron Garrison inspired us with his gifted color photography reflecting his appreciation of the Bluegrass landscape. Jim Rebmann painstakingly printed the black-and-white photographs (which were taken by the authors unless otherwise indicated). Gyula Pauer and Marvin Rinck drafted the maps. R. Matthew Wooley and L. Martin Perry read chapters two and four respectively and made helpful suggestions. Julie Riesenweber of the Kentucky Heritage Council provided careful and insightful editing and improved the manuscript with valuable recommendations.

Introduction

> If there are ever sermons in stones, it is when they are built into a stone-wall,—turning your hindrances into helps, shielding your crops behind the obstacles to your husbandary [sic], making the enemies of the plow stand guard over its products.
>
> —John Burroughs, *Signs and Sermons*

While observers of the American countryside may view rocks as obstacles, hindrances, or enemies, as the quote above implies, in the Kentucky Bluegrass, rocks are a blessing. Ancient limestones yielded the region's fertile soils that provided the basis for the luxuriant vegetation so admired by both native American Indians and European explorers. These same limestones were, with effort and expense, quarried to provide fence material.

Two themes guide this study of central Kentucky's rock fences. First, these fences are a significant part of the state's distinctive Bluegrass landscape created by the interplay of the physical environment, culture, and technology over the past two centuries. The land's qualities—fertility, vegetation, drainage, and geology—have provided the physical context for inhabitants' ability to make a living and fulfill their expectations. The limestone bedrock upon which the Bluegrass lies and of which its rock fences are constructed is the physical feature that has been most influential in the region's settlement and subsequent development. Second, the Bluegrass landscape did not develop in isolation but was linked to other places in America and abroad. Some connections were tangible, such as the routes along which people traveled and the people with whom they associated. Other linkages were less direct; traditions and habits brought from other places influenced how people used the land and what they built on it. The flow of ideas from distant places by way of newspapers, books, and letters augmented existing customs.

By the 1880s, quarried rock fences were the most common fencing type on Inner Bluegrass farms. This subregion of the greater Bluegrass (fig. I.1) is centered around Lexington and extends north-south from Cynthiana to Danville and east-west from Lawrenceburg to Winchester. A second subregion, the Eden Shale, surrounds the Inner Bluegrass as a ring of stream-dissected hills. Rock fences were common here as well, and many remain.

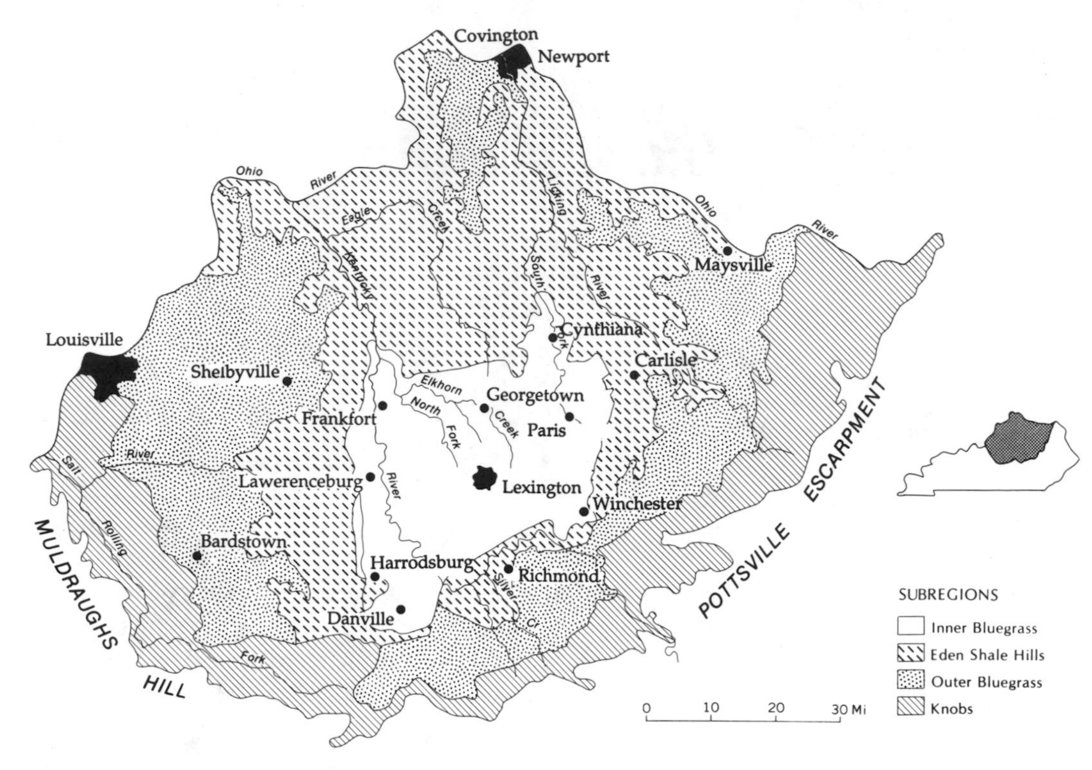

Fig. I.1 SUBREGIONS OF KENTUCKY'S BLUEGRASS REGION

Such fences also survive in the Outer Bluegrass counties of Bracken, Mason, Nelson, and Washington, and in the Pennyrile counties of Green and Adair, outside the Bluegrass (fig. I.2). This study, however, concentrates geographically on the Inner Bluegrass and Eden Shale subregions, areas once dominated by extensive rock fence networks.

Questions about both the rationale for rock fence building and the identities of the builders persist, in part because the oral traditions of fence masonry are largely obscured. Few masons working today know the background of the "rock fencers" who preceded them. Nor do many people understand the structural subtleties of fence construction.

In the absence of accurate, readily available information about the region's rock fences, inaccurate popular perceptions about them have been treated as fact. One prevalent notion is that slaves built the fences on antebellum plantations during the winter season when farm activity slowed. This idea, gaining strength when it found its way into print, recently appeared on the front page of the *Wall Street Journal* in an article on shopping malls in the Bluegrass: "Stone fences built by slaves line the roads that twist through rolling fields where mares nuzzle their foals in the sweet grass" (Swasy 1989). Once published, such information reinforces popular beliefs,

Fig. I.2 COUNTIES IN KENTUCKY'S BLUEGRASS REGION

thereby perpetuating false perceptions. Evidence in this study contradicts many commonly accepted ideas about central Kentucky's rock fences.

An additional question is why people built rock fences in Kentucky when split rail fencing was common in surrounding areas. One popular explanation is that farmers built rock fences to use the fieldstone cleared from their land. While this is true in the Eden Shale area, it is not true for most of the Inner Bluegrass.

A third issue concerns the origin of Bluegrass fence forms. What is the relationship between fence form and fence masons? Do all fence masons follow the same construction plans? Stone masonry is not a trade easily mastered by untrained builders. Whose traditions do the fences represent?

Scholarly research on fence building is informative, although the quantity of such work is small. In a 1954 trip through Georgia, North Carolina, Virginia, Tennessee, Kentucky, and Ohio, geographers Eugene Cotton Mather and John Fraser Hart found rock fences only on estates in Kentucky's Bluegrass region, where rock was equal to woven wire in frequency of occurrence. Another study by geographer Wilbur Zelinsky questioned the rationale for construction of walls and fences in the eastern United States (1959). Why, he wondered, were rock fences common in New England, whereas stone was rarely used for house walls? Why, in Pennsylvania, Maryland, and Virginia, was stone an important house fabric yet seldom used for fencing? The Bluegrass offers an exception to Zelinsky's contention that stone houses and rock fences do not occur in the same locales, since the landscape contains many buildings constructed of the same limestone as its fences.

Other writers associate rock fences with regions of thin, infertile soil. While this observation holds for central Wisconsin and the New England hills, as well as the Eden Shale region of Kentucky, it is not true in the Texas Hill Country, where nineteenth-century German farmers built hundreds of miles of quarried rock walls (Jordan 1966), nor in Kentucky's Inner Bluegrass. Whether the fences are found in New England, Wisconsin, Texas, or Kentucky, there is little argument, however, with H.F. Raup's statement that fences are a "significant element of material culture, indicative of physical surroundings, having special social significance, and acquiring different forms which may stem from culture contact and tradition" (1947, 7). Bluegrass rock fences reflect both landscaping traditions of landowners and fence-building customs of area craftsmen.

Information for this book came from diverse sources. We compared historic travelers' descriptions of the region with farm records. Manuscript collections supplied valuable data. Deed research established dates of fence construction on particular farms. Federal census manuscripts produced names, occupations, and birthplaces of nineteenth-century masons. Farm owners discussed fence construction on their properties. Interviews with masons, both active and retired, provided terms for the glossary and descriptions of fence construction. Masons also explained how they learned their craft. Newspaper reports, both historic and contemporary, corroborated other information.

Since these sources are ambiguous in their use of various terms, an explanation about this book's terminology is in order: there is a difference between a rock fence and a stone fence. Strictly speaking, a rock fence is built of rocks gathered from fields and creek bottoms, while a stone fence is built of material that has been quarried and shaped, or dressed. A mining executive explains the difference in these terms: rock is the material that is in the ground and that comes from the ground, whereas stone is something that has been made of the rock. "The term stone has no scientific standing . . . but the word does have a proper place when the commercial usage of earth materials is referred to" (Cassidy 1977). The majority of fences in the Bluegrass were built of undressed quarried rock,which is more like rock than like stone. Because of this blurring of definition, and with respect for the local tradition of calling either kind of fence a "rock fence," the term "rock fence" is used here, except when quoting documents that use the word "stone."

Such fences are termed "stone walls" in New England, "stone rows" in northern New Jersey, "stone fences" in Pennsylvania and western New York, and from West Virginia south, "rock fences" (Meredith 1951, 125), so that terminology in America is only slightly less confusing than in the British Isles where rock fences also exist. In Scotland and the north of England, fences of rock are called "dykes" or "dry walls." In Northern Ireland, they are "ditches" (Robinson 1989) and in Cornwall, "hedges" (Hart 1980, 89). The term "fence" is of long standing, however, in the British Isles. The word "fence" was used in legal records of 1647 and 1680 to indicate rock walls in Bordley and Settle, villages in England's Yorkshire Dales (Raistrick [1946] 1988, 26-27).

In explaining historic fence masonry practices and describing existing rock fence features, we made choices about verb tenses: If a practice occurred only in the past, we explain it in past tense. If a practice began in the past and continues to the present, we explain it in present tense. If a historic fence contains features that were constructed in the past but that remain in the existing fence, we describe them in present tense.

Because many fence features are known by more than one name, the glossary provides fence terminology definitions. It also indicates places where various terms are used and thus provides linguistic evidence of the link between Kentucky fences and their builders' traditions. Appendix 1 contains the names of stonemasons listed in the U.S. Federal Census manuscripts for several Bluegrass counties; it also provides the names of some contemporary masons. This list gives the masons' birthplaces and makes possible tracing the work and movements of individual masons.

This book serves to dispel a common myth about rock fences, one of the region's most venerable symbols. The myth has masked a complex process that involved migrating masons who brought traditional skills to the Bluegrass and farmers who responded to legal and practical fencing requirements by utilizing local materials. If this work inspires stewardship of the rock fences of the Bluegrass, we are well repaid.

Ron Garrison

CHAPTER 1

Landforms, Rock, and Quarrying

Only in the hallucination of some strange form of insanity could he believe that Kentucky is not the fairest land of the Creator, and the Bluegrass region its paradise.

—Basil W. Duke, *Reminiscences*

Kentucky's central limestone plain, known for two centuries as the Bluegrass,[1] is not uniform in topography or opportunity but is comprised of three distinctly different subregions, each having a different landscape—the Inner Bluegrass, the Eden Shale hills, and the Outer Bluegrass. The varying characteristics of each subregion had a significant effect upon the way settlers used the land and the structures they built (cf. Mead 1966). Differences in the predominantly limestone bedrock demarcate these subregions. Similarly, the varied bedrocks became material for different fences in each subregion.

Impressions of the Bluegrass

The gently rolling land, deep fertile soil, and lush vegetation of the Inner Bluegrass, the centermost subregion, attracted settlers beginning in the last quarter of the eighteenth century. Some of the best accounts of the area's extraordinary bounty are those of travelers from Europe and the eastern seaboard who often described it as an Eden-in-the-West. In 1834, for example, an anonymous but articulate traveler observed that as he approached Lexington, near the center of the region,

> the road winds through an open champaign country of the most attractive aspect. The surface is not broken by hills, nor is it level—but of that beautifully rolling or undulating character, which is above all others the most pleasing to the eye, and the best adapted to the purposes of husbandry. . . . The soil is of the richest kind, and the improvements superior to any that I have seen in any part of the United States. I had long been aware of the high character claimed for the country around Lexington; but, prepared as I was to behold a region rich in attractive scenery, . . . I

was agreeably surprised in finding that it surpassed my anticipations. [Schwaab 1973, 266]

At the time of exploration in the eighteenth century, the Inner Bluegrass was covered by cane lands and open hardwood forest that the area's farmers eventually converted into woodland pastures for cattle. Estate owners created these semishaded parks by removing the smaller trees from the virgin woods, leaving the largest trees spaced sufficiently to admit sunlight so that grass would grow.[2] Englishman James Silk Buckingham admired these pastures in 1842 and wrote that they "form delicious retreats for the cattle that are fed there, and look as beautiful as most of our English parks" (1842, 507). Such favorable comparison of the Inner Bluegrass with the English landscape was a recurring theme over the next century and a half.

The second Bluegrass subregion, the Eden Shale hills, surrounds the Inner Bluegrass. The landscape here changes form abruptly where streams have cut steep-sloped valleys into the surface, and the land is less productive than that around Lexington. Visiting Kentucky in 1818, traveler James Flint noted this change. Some twenty-five miles from Lexington, traveling northeast toward Maysville, he found the land to be "hilly, poor, and for the most part covered over with detached pieces of limestone. . . . Rude implements are left to rot in the field; and the scythe allowed to hang on a tree from one season to another. . . . Economical agriculture has no place here" (Thwaites 1904, 146-47). New Yorker Charles Hoffman noticed similar conditions in 1835 on his way south from the Ohio River. North of Georgetown in Scott County, the land was "broken up into undulations so short and frequent as to make the office of ploughing the hill-sides no sinecure" (1835, 121). Though not lettered in earth sciences, these men were noting the changes in topography and economy that accompany changes in the underlying bedrock. Some central Kentucky rock strata produced a surface that was highly valued for farming, and others did not.

The Outer Bluegrass, the third regional subdivision, lies outside the ring of Eden Shale. This landscape resembles the Inner Bluegrass, although the soils are not as deep and are somewhat less fertile. John Burroughs traveled the area in 1893 and acknowledged with a flourish the role of the underlying limestone in this locale: "Thus, upon this lower silurian, the earth that saw and nourished monsters and dragons was growing delicate blue-grass. . . . I thought I had never seen fields and low hills look so soft in the twilight; they seemed clad in greenish-gray fur" (1895, 224).

Although the differences between the three subregions are obvious, travel writers and the popular American agricultural journals of the middle 1800s often glowingly described the Bluegrass landscape as an aggregate. A New York publication, *American Agriculturist,* for example, informed readers that the Bluegrass "may be pronounced one of the most fertile and eligible agricultural districts upon the globe" (1842, 67). Significantly, the editor also reported that many of the beautiful plantations were strongly fenced with high stone walls (139-40). The entire region gained fame for its aesthetic qualities, and its rock fences played a role in acquiring that reputation.

Kentucky author James Lane Allen wrote a complete book attempting to "portray adequately" his feeling for the Bluegrass. Critics may find his writing oversimplified romanticism, but, as this passage reveals, Allen understood that the essence of the relationship between the Bluegrass countryside and its people resided in the area's physical geography, especially the stone below the surface: "One cannot sojourn long without coming to conceive an interest in this limestone, and loving to meet its rich warm hues on the landscape. It has made a deal of history: limestone bluegrass, limestone water, limestone roads, limestone fences, limestone bridges and arches, . . . limestone water-mills, limestone spring-houses and homesteads—limestone Kentuckians! Outside of Scripture no people was ever so founded on a rock" (1892, 26).

Admiration of the Bluegrass landscape did not diminish as years passed. In the 1920s geographer Darrell Davis recorded impressions that vary from those of a century earlier only in detail: "The Blue Grass can lay legitimate claim to possessing a certain type of attraction second to that of no area of equal size in North America. . . . Pastures and fields are gently rolling and are frequently almost grove-like in character, so numerous are the trees. . . . Stone fences flank these roads for miles [and] furnish a rural setting which is difficult to portray adequately" (1927, 52).

Rock and Regional Variation

Changes in the bedrock across the Bluegrass account for the subregional differences that travelers and journalists observed. Through millions of years, geological forces uplifted the rock beds of an ancient sea floor, forming a shallow dome centered in the Inner Bluegrass. Over the aeons, the region's streams and rivers eroded the top off the dome. The result is much like an onion with the top third sliced off: the inner layers appear as circular rings. Rock layers exposed near the region's center are oldest, and they become progressively younger in all directions. The oldest exposed rocks in Kentucky—Middle Ordovician limestones—underlie the soils of the Inner Bluegrass as a consequence of this geology. The various rock beds throughout the region are not the same in chemical and physical composition but vary substantially from layer to layer, these variations affecting soil fertility and erodibility (fig. 1.1).

Through the constant weathering effects of water and frost, limestone

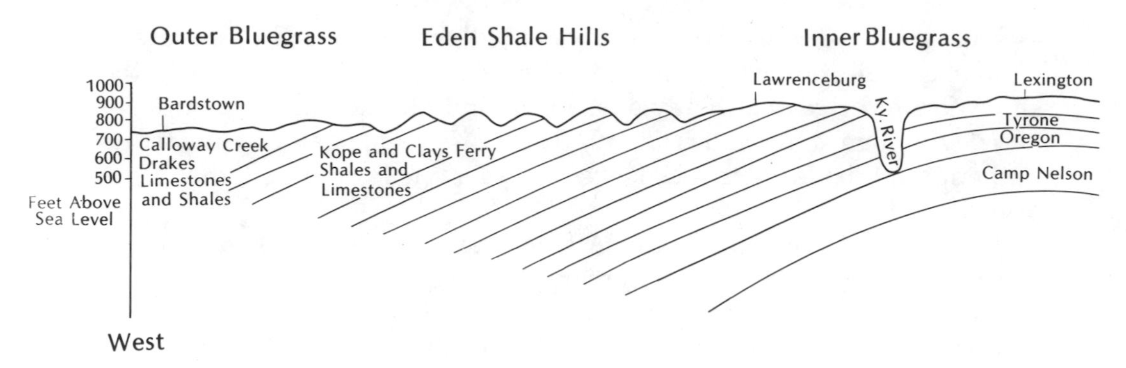

Fig. 1.1 CENTRAL KENTUCKY TOPOGRAPHIC AND GEOLOGIC CROSS SECTION

becomes soil. In the mid-1880s, Dr. Robert Peter tested soil and rock samples across the Inner Bluegrass in search of a scientific explanation for the region's fertility and perpetual productivity. He ventured an interpretation as to how minerals locked in the limestone were made available to plants: "The atmospheric water, falling as rain . . . always brings with it, in solution, a certain quantity of carbonic acid. . . . Penetrating the soil [and] coming in contact with the limestone rock; this acidulated watery solution continually dissolves out [into the soil], not only the carbonate of lime . . . but also carbonate of magnesia and phosphate and sulphate of lime, and small proportions of the alkalies—potash and soda—which it always contains" (Perrin 1882b, 16-17). Peter determined that soil quality depends on the bedrock's mineral content.

The bedrock surrounding Lexington—known as the Lexington Limestones—yields through this dissolving action soils that are vastly superior for agriculture to those outside the area. One member of the Lexington Limestones, Tanglewood, for example, is prized bedrock as far as the region's farmers are concerned because it produces the deep Maury-McAfee soil whose high phosphate content causes it to be greatly valued. This deep, cinnamon-colored soil is among the most fertile in the South and has sustained planting and grazing for two hundred years. Usually buried under two or more yards of soil, rock is rarely found on the surface even after years of tillage, except along stream valleys or the sides of sinkholes.

These types of bedrock limestone also provide excellent building material. A St. Louis periodical reported that "there are few countries where stone are so perfectly available for building good walls as in the better portions of Kentucky. . . . Limestone is found in the best possible form for building stone wall at the least possible cost. The stone is found in the quarries in layers of from two to six inches thick, with alternate layers of clay; this stone is easily quarried and broken into pieces of suitable size for building stone wall" (*Valley Farmer* 1857, 42-43).

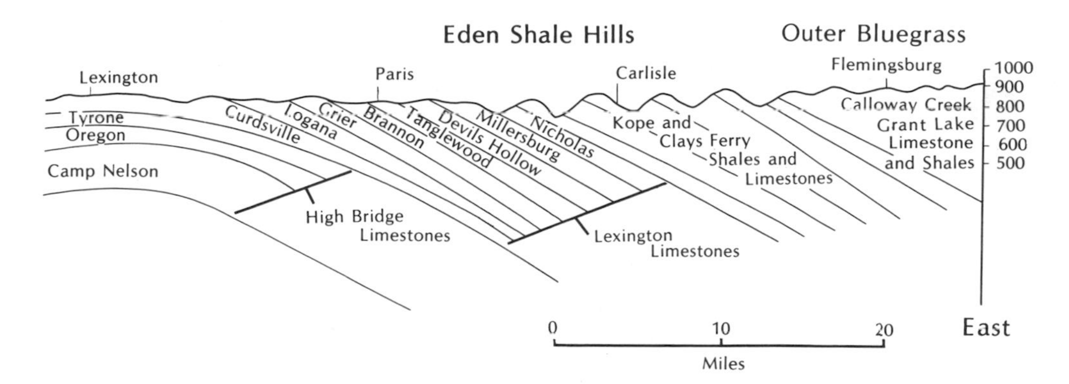

Fig. 1.1 *Continued*

Two rock types prized as superior construction stone are the Oregon and Tyrone limestones. These beds formed below the bedrock around Lexington, the Lexington Limestones, but are exposed by erosion in the Kentucky River gorge and its tributary creeks. Oregon is a golden-colored freestone, having no dominant grain, that is commonly called "Kentucky River marble." Tyrone, nicknamed "birdseye limestone" because of its small calcite facets, is very dense and has smooth glass-like surfaces that weather to a bright white. Until the early twentieth century, Oregon and Tyrone rock appeared in fences only in areas where they outcrop, although both were quarried for use as dimension (cut) stone. Rock companies produced Tyrone commercially and hauled it to Lexington and surrounding horse farms where it became a popular material for the decorative fences built both before and after the turn of the twentieth century as well as for buildings. Where this rock was used in fence construction, its white, sharp-edged block shape is easily recognizable (McGrain 1982; Richardson 1923).

The three hundred feet of limestone beds that lie above the Oregon and Tryone, collectively known as the Lexington Limestones, vary in thickness, color, texture, and chemical content (McDowell, Grabowski, and Moore 1981). Some formations in this series contain fossils and carbonates. Their colors include all shades of gray as well as blue, gold, and brown. The stone at Cane Ridge in Bourbon County, for example, is conspicuous for its gold and rust colors, while some beds of the Cynthiana formation are sea blue. All Lexington Limestone members were locally quarried and used in building construction between 1785 and 1835; they were quarried and used for rock fences into the mid-twentieth century (Black, Cressman, and MacQuown 1965; Murray-Wooley 1985).

Twenty miles or more in all directions encircling Lexington, the Eden Shale belt contains steep terrain underlain by limestone strata of the Late Ordovician period interspersed with thick shale beds. The shales weather into yellowish clay soils that wash away in heavy rain or bake to a brick-like

Fig. 1.2 LEDGE QUARRY. Five quarries on this Scott County farm, last used in the mid-1800s, are near existing rock fences that border pastures and barn lots.

consistency under the summer sun. Few plants flourish in the Faywood-Eden soils that form in this subregion where fertility has declined over the past century or more as farmers cleared the oaks and beeches from the hillsides to grow crops. Once cleared, what little soil clung to the steep slopes eroded into the valley bottoms after only a few growing seasons.

Rampant soil erosion, both natural and that augmented by humans, left the Eden Shale hillsides littered in rock since limestones separating the shale beds are much more resistant to weathering than is shale. It is common for limestone rocks to occupy nearly as much of the surface as the grass in a hillside pasture. Each time the steep fields are plowed, each time it rains, more rock appears. Using this ready supply of field rock for fencing material was an obviously practical choice for farmers in these hills.

The third Bluegrass subregion, the Outer Bluegrass, is underlain by a circular band of the area's youngest rocks: Upper Ordovician, Silurian, and Devonian. Bedrock is predominantly limestone, and the land surface is gently rolling. Although the soil is much more productive than that of the Eden Shale, the region did not foster as many large and prosperous landholdings as the Inner Bluegrass, and it has fewer rock fences.

Most rocks in the Outer Bluegrass occur as distinctive individual beds. Laurel Dolomite, which lies near the surface in eastern Jefferson, Oldham, Bullitt, and Nelson counties, is an excellent dimension stone, and masons used it for the construction of buildings in the early 1800s (Murray-Wooley 1987). Its use for rock fences in these counties, however, is rare. Much more common in the Outer Bluegrass are fences primarily built of creek rock or ledge-quarried limestones. While these are scattered throughout Outer Bluegrass counties such as Adair, Green, and Washington, only in Garrard and Mason counties are such fences numerous.

Obtaining the Rock

Ledge and Creek Quarries. Fence builders across the Inner Bluegrass, where bedrock lay well below the surface, searched for places where they could obtain rock without removing considerable amounts of earth. Surface erosion at sinkholes, springs, and along creek banks sometimes exposed the underlying rock strata and formed rock outcroppings or ledges. If the quality of the rock at these places was suitable for construction, it could be readily extracted. Quarriers obtained the bulk of rock for the fences of the Inner Bluegrass from numerous such hillside ledge quarries (fig. 1.2).

The top layers in a ledge quarry are called overburden or "ledge rock." Because exposure to water from the surface subjects this rock to the chemical decomposition process noted by Dr. Peter, it is usually pocked and uneven. This dissolving effect is most pronounced on rock in the uppermost beds, giving it an uneven character compared to rock deeper in the quarry, which breaks into more uniform flat-surfaced slabs. Because of its unevenness, stonemasons do not use ledge rock in the construction of buildings (Kelly 1987). It is good fence material, nonetheless, and was widely used for this purpose throughout the Bluegrass region. Ledge rock fences have irregular layering or coursing because the ledges vary in thickness and each rock had a unique size and shape (fig. 1.3). This rock is, however, usually more uniform in shape than field rock.

Fig. 1.3 LEDGE ROCK FENCE. Different ledge thicknesses are reflected in varying rock sizes in this Scott County fence.

Fence builders used wedges and sledgehammers to crack off long rock slabs from exposed ledges. Other necessary tools included levers, picks, and crowbars. For economy in hauling, workmen opened ledge quarries as close as possible to the fence-building site and uphill from the section under construction. Draft animals pulled sleds loaded with rock from the quarry to the fence. One informant, who as a boy assisted his father in building fences, described these rock sleds as having runners sheathed in sections of wagon wheel rims (J. Huff 1989).[3]

Because quarries were located as near fence sites as possible, the pattern of relict quarries today evidences the location of old fencing, even if the fences built from their rocks are gone. Landowners often directed that quarries be filled with soil after use to level the land for cultivation or to prevent falls by livestock. On some Inner Bluegrass farms evenly spaced shallow depressions in pastures are old ledge quarries that provided rock for nearby fences. On other farms, the quarries were centrally located and much larger and deeper. Many are still open, surrounded by trees and guarded by fences. Those that were deep enough to intersect underground drainageways filled with water, and farmers often use them for stock watering ponds.

The beds and sides of flat-bottomed creeks are also convenient places to extract rock. The shapes and surfaces of rock from creek sides are similar to ledge rock. Rock prized from creek beds (fig. 1.4), in contrast, is slab-like,

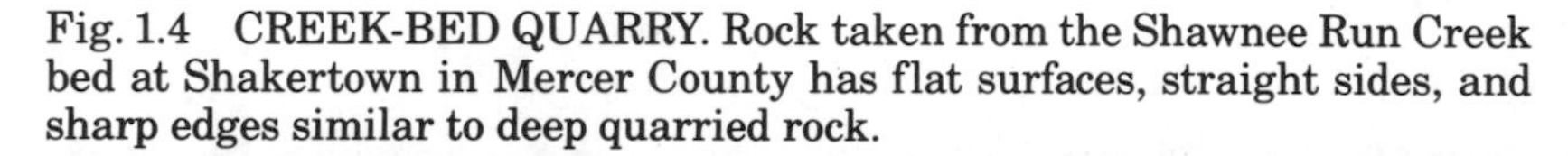
Fig. 1.4 CREEK-BED QUARRY. Rock taken from the Shawnee Run Creek bed at Shakertown in Mercer County has flat surfaces, straight sides, and sharp edges similar to deep quarried rock.

with smooth tops and square edges and a uniformity so pronounced that it is difficult to distinguish from deep-quarried rock.

Deep-Quarried Rock. Some farmers quarried fence rock themselves while others utilized slaves or hired hands for this work. Wealthy planters, however, hired quarrymen. Nineteenth-century census takers often listed "quarryman" as an occupation distinct from "stonemason" or "fence mason." The landowner rather than the mason typically hired the quarrymen to quarry and haul material for construction projects such as roads, building foundations, and fences. For some jobs the quarrier and the mason were the same person, in which case the price for the fence included the price for quarrying. An 1857 farm journal reported, for example, that "the work is usually done by contract, the laborer quarrying the stone and laying up the wall" (*Valley Farmer* 1857, 42). Bourbon County plantation owner Brutus J. Clay paid his fence builder in 1848 for quarrying rock and building fences but had his slaves haul the rock from the quarry to the building site.

Professional quarriers discarded the weathered overburden near the surface to reach the deeper beds. Unweathered limestone is usually quite sound and lies in relatively uniform layers (figs. 1.5 and 1.6). Deeper quarries provided true construction stone that was used on the fertile Inner Bluegrass plantations for high quality ashlar fences as well as for buildings. Ashlar fences are distinctive because their material has square edges like cut stone or brick. The blocks from deep quarries are more uniform than ledge rock, with horizontal faces that are comparatively flat and that break cleanly to form right-angled vertical sides. The sharp, clean edges, faces, and corners allowed the mason to build fences having uniform horizontal courses with tightly fitting joints (fig. 1.7). Ashlar fences built of deeper quarried rock are properly called "stone fences" because the mason shapes the stone with stone hammers and chisels. Still, most Kentuckians refer to them as "rock fences."

An observant nineteenth-century farm journal editor distinguished between weathered rock and that of deeper layers: "Quarry stone usually make better walls. . . . They are more shapeable, with flat surfaces, lie more compactly and evenly, . . . and where not too expensive, even if surface [field or ledge] stones are at hand, are to be preferred. Yet surface and quarry stones do not always abound on the same premises, and the farmer is in most cases compelled to choose either one or the other. [The fences in central Kentucky] are of quarried stone chiefly . . . and are built in the most substantial manner" (*American Agriculturist* 1859, 110).

Quarrymen commonly used blasting powder, or black powder, to remove deeper rock from quarries. Farm ledgers of the period contain receipts for blasting powder by the keg costing from $5.00 each in 1848 to $3.75 in 1855. Priming powder cost $.38 per pound and safety fuse $.75 per hundred feet (Rion and Mitchell 1848; Wilgus and Bruce 1855; Rion and Sharrard

Fig. 1.5 ROAD CUT. The top third of the roadcut is loosely bedded ledge rock having weathered edges, while lower bedding planes are so tight they are nearly indiscernible. Vertical lines are power-drilled holes for blasting.

Fig. 1.6 QUARRY FLOOR. Although the quarry bed is here partially covered with fallen leaves, its flatness is apparent in this Tyrone Limestone quarry near Boone Creek in Fayette County.

Fig. 1.7 QUARRIED ROCK FENCE. Block-shaped quarried rock permitted level coursing, flat faces, and tight fitting joints in this well-built fence in Bourbon County.

1851; Mitchell 1859; Parrish 1855). According to Stanley Kelly, a Mercer County stonemason and restoration specialist, "black powder is the only explosive that can be used in quarrying building stone. Dynamite shatters the stone. . . . Black powder has a pushing effect and is used mostly for removing the overburden" (1987). Quarriers used blasting powder to extract rock for fences into the present century. The Lee family masons, for example, blasted rock at an old quarry on the Alexander farm in Woodford County until about 1910 (Cause 1980). Frank McGarvey was still blasting rock for fences in Woodford County in the 1940s (Gormley 1987).

Bore holes and tool marks in old quarries indicate the technique used to break the rock away from the quarry face (fig. 1.8). Quarriers used steel-tipped drills of about one-inch diameter to drill holes into the top of the block to be removed. The drill was simply a shaft with a flattened end that had been beveled to a sharp edge. While one man held the drill, another hammered it against the flat bedrock surface. By rotating the drill as it was hit, quarrymen could bore holes into the hard limestone rock.[4] They drilled holes all along the intended fracture line and, when the holes were deep enough, filled them with blasting powder. The charge pushed the rock out of place; it could then be further broken up if necessary. Extant blacksmith bills to a landowner include charges for making, repairing, and sharpening quarrymen's tools: for making a blowing needle, dressing drill points, steel-

Fig. 1.8 BORE HOLES. Now festooned with ferns and mosses, this old ledge quarry bore hole is one in a line of hand-drilled holes that, when filled with powder and blasted, broke away the quarry face.

ing and dressing picks, facing a sledgehammer, mending a pick, and sharpening shovels (Bealeret 1851a, 1851b).

If the quarry was excavated on nearly level land, a ramp provided access to the quarry floor. Laborers drove wagons or rock sleds pulled by horses, oxen, or mules close to the working face of the quarry, where they loaded rock for hauling to the building site.

Fig. 1.9 ROCK DUMP. The swale in the foreground contains two rock dumps and the opposite hillside another, in the Eden Shale region of Mercer County. A rock fence borders the wet-weather creek in the valley; others appear as dark lines across the fields.

Field Rock. There have never been quarries in the rugged Eden Shale hills other than those opened to provide material for road work. Instead, farmers constructed most rock fences in this area with rock gathered from the fields. "Pick-up rock" was, and continues to be, readily available for fence construction in the Eden Shale subregion. Each cultivation season brings more rock to the surface, and farmers remove it to facilitate tilling and to provide more space for grass. While some rock required pry bars and levers to free it from the soil, much lay loose on the ground.

Clearing rock from the fields is prerequisite to farming in these hills, and farmers faced the problem of how to discard it. Field rock is not suitable for construction of buildings since, having been exposed to the weather, it has "rotted" or deteriorated and cannot be shaped without breaking or crumbling. Some farmers simply loaded the rock on sleds, hauled it to the center or edge of the field and piled it there. Others hauled their rock to the nearest ravine or drain where they piled it into broad dams to retard the speed of rainwater as it washed from the fields (fig. 1.9). Some saw that the volume of

Fig. 1.10 FIELD ROCK FENCE. A field rock fence bordering pastures in the southeastern Bourbon County exhibits typically weather-rounded field rock having a smoother shape than ledge rock (fig. 1.3) and far less uniformity than quarried rock (fig. 1.7)

rock was much greater than these simple solutions accommodated. These farmers collected rock where they needed enclosures and laid it into fences, often increasing the fence widths as necessary to use up the unwanted material.

Fences built of field rock are easily identified. The rock is usually heavily weathered, causing it to be irregular in size and shape, with rough, pockmarked faces and tapered, pointed edges. It usually has no flat face and when broken in half is lens-shaped in cross section (fig. 1.10).

Creek Rock. An additional source of material for fence construction in all Bluegrass subregions is loose, water-worn rock from creek beds. This rock is much different in texture and shape from ledge, quarried, or field rock or from its near neighbor, rock prized from creek beds. Loose creek rock has smooth surfaces and rounded edges resulting from the tumbling action of rock against rock in the watercourse (fig. 1.11). Fences built of creek rock are relatively rare, in part because such stone is not in great supply in central Kentucky. Broad streams with accessible flood plains, such as the Salt River,

Fig. 1.11 WATER-WORN ROCK. Rocks smoothed by tumbling creekwater are easily recognized by their soft texture when interspersed with field rock in a fence. The lens-shaped rock in the second row is a flat, rounded creek rock broken in half and laid in the fence with its broken face outward. Several coping members also are creek rock.

are few and far apart, and smaller creeks with shallow channels have limited erosive power. More important, masons dislike creek rock for fences because it is too smooth and rounded to have gripping power. Creek rock, therefore, when used in fences is interspersed with ledge and field rock.

The type of material used in fence building—whether field rock, ledge rock, creek-bed rock, deep-quarried rock, or water-worn rock—makes a difference in the structure as well as the appearance of the fence. Fence forms also differ over time and place. Some of these differences are obvious, logical, and self-explanatory; others require an understanding of how rock fences are constructed.

CHAPTER 2

Rock Fence Construction

In no country have we ever seen better stone walls than are now built in Kentucky.

—*Valley Farmer,* 1857

Like many things people place on the landscape, rock fences seem to be simple structures: rocks piled at the edge of a field to form a barrier that confines stock. Close examination, however, reveals that the fences are complex and employ in subtle ways the physics of friction, angle, and gravity to maintain cohesion and stance.

Rock fences in Kentucky have few common characteristics with those in New England, another region of America where they are prolific. Although fences in both places are built without mortar (dry-laid) of locally obtained rock, the similarities end there. Even within Kentucky, details of coursing, bonding, foundations, and copings vary from county to county, sometimes between locales within a county. Kentucky's fence forms also changed considerably over time. Some differences in structure are the results of stonemasons' adaptations to local variations in rock types, slope angles, or soil stability. Local techniques also developed when apprentices learned and continued the methods of senior craftsmen.[1] Other differences derive from cultural legacies, individual workmanship, and changes in fashion and technology.

The Character of Extant Fences in Kentucky

Rock fences in the Bluegrass fall into two major categories, dry-laid and mortared. Each category has various subtypes: dry-laid fences are flat-coursed or vertically coursed and are of the late eighteenth-century plantation, nineteenth-century turnpike, or twentieth-century modern eras. Mortared walls may have concrete block or rock and concrete cores and may be faced with stone in a variety of patterns. Descriptions in this book of each fence type and subtype and of specific construction features of typical examples explain how and why the fences differ. The Glossary defines terminology used in fence descriptions and gives synonyms for the terms, identifying places where the synonyms are used.

The dry-laid flat-coursed fence is the most common rock fence type in central Kentucky. The two principal divisions of this type, although similar in appearance, do not have the same physical structure. The oldest of these—built between the 1770s and the first half of the 1800s—is called the "plantation fence" in this book.[2] The duration of the second dry-laid fence subtype has some overlap with that of the first. It was built during the last seven decades of the nineteenth century but did not become dominant until mid-century. Since this fence type became ubiquitous during the turnpike construction period of the mid-1800s, these fences are identified here as "turnpike fences." Although there is a structural blending of these two subtypes in the mid-1800s, as well as an overlap in dates, the two are distinguished to facilitate discussion. A third dry-laid fence subtype, the "edge fence," is not flat-coursed but has upright diagonal courses the full height of the fence. These fences overlap the construction period of turnpike fences; the oldest date to the mid-1800s and the newest to the present.

The mortared wall, the second major rock fence category, became popular around the turn of the twentieth century and is the primary type built today. Subtypes of mortared fences include those with a concrete and rubble core and those with a concrete block core. Since the use of mortar allows a mason artistic leeway in assembling the fence facade, mortared fences have designs including patterned and ashlar walls as well as untooled quarried, field, and creek rock faces. The craft of building with mortar is very different from the craft of building "dry," so much so that the two skills are related only in use of the same basic material. The reason for discussing two such different trades in the same work is that residents of and visitors to Kentucky's Bluegrass seldom differentiate between a dry-laid structure and a mortared one; both are casually termed "rock fences" and both are little understood.

Both mortared and dry-laid fences are frequent in the Inner and Outer Bluegrass and remain in some places in the Pennyrile region of south-central Kentucky and the Appalachian Mountains of eastern Kentucky.

Dry-Laid Rock Fences

Plantation Fences. Scattered references to the plantation-era fences occur as early as 1777 (Draper 12C:26-29), but these fences did not come into wide use until the early nineteenth century. Plantation fences originally surrounded barnyards, stockyards, paddocks, house yards, graveyards, gardens, pastures, and fields; some are boundary (or line) fences between landowners (figs. 2.1-2.3). The greatest remaining number enclose graveyards, since families often retained burial grounds when farms changed hands and the

Fig. 2.1 BARNYARD FENCES. This Boyle County barnyard includes a tobacco and horse barn. A second tobacco barn is in the background.

fences were therefore not affected by field enlargements or road improvements.

Plantation fences are of quarried rock, creek rock, or field rock. Regardless of rock type, these dry-laid flat-coursed fences have common structural elements: double-wall construction, tie-rocks, battered sides, and solid cap courses or full-width coping rocks. Fence strength and durability is the result of skillful placement in which each rock is locked securely in place. The rocks are only "attached" to one another by pressure, so that binding power depends on weight, surface friction, and gravity. While the minor details of plantation fences vary, most early nineteenth-century examples are similar in form, if not identical in dimensions (fig. 2.4).

In order to construct this fence type, the mason or his helper[3] dug a trench in which to build the foundation. This trench was usually only four to six inches deep, extending to solid subsoil. A constructed foundation was unnecessary in areas of shallow soil, where the builder could lay the bottom

Fig. 2.2 PASTURE FENCES in Boyle County.

Fig. 2.3 BOUNDARY FENCE in Bourbon County.

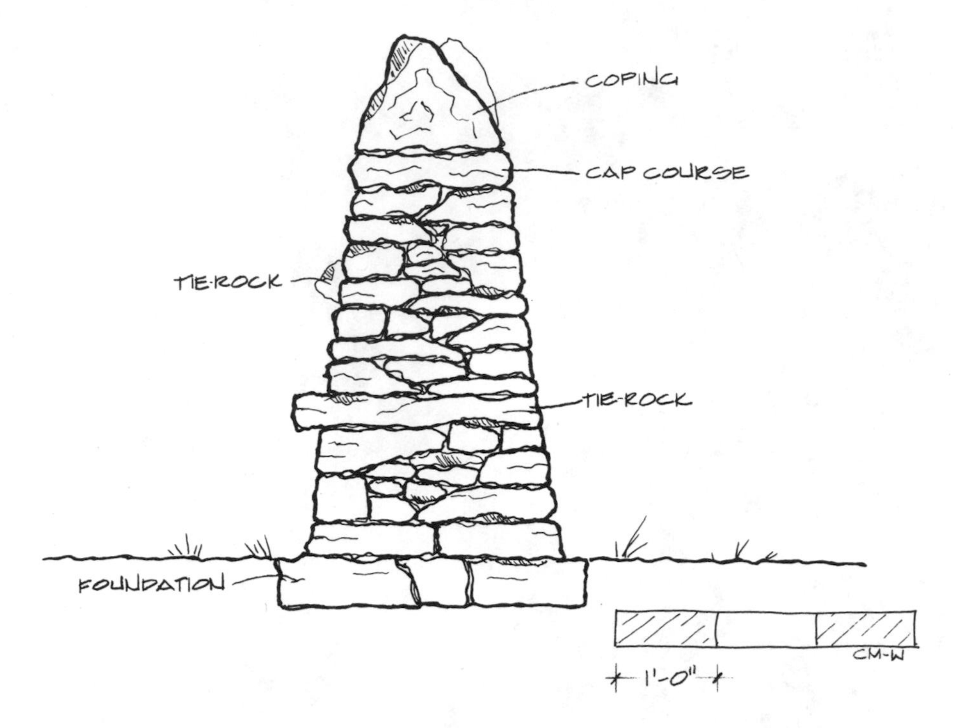

Fig. 2.4 SECTIONAL DRAWING OF A PLANTATION FENCE

course of the fence directly on bedrock. Some fences do not have a foundation but commence at ground level with only the sod layer removed. One mason cautions that these fences should only be built during a "light [new] moon"; if laid "during a dark [full] moon it will sink in the ground and fall down" (Kelly 1989b). Here, a new moon is waxing or becoming lighter, and a full moon is waning or becoming darker.

Where the intended fence line went down a hillside, workmen dug a trench in level steps so that the mason could always lay the foundation stones on a true horizontal. If the fence ran across a hill, the foundation was broader on the downhill side. In either case, masons placed foundation rocks directly in the trench to span a width four to eight inches greater than the fence base and stacked them one or two courses high to raise the top of the foundation to surrounding ground level. The mason used the largest available rocks and arranged them to cover the full width of the foundation solidly, with the largest rocks forming straight lines on the outside edges.

The builder began constructing the fence itself at ground level upon the foundation, placing the lowest rocks two to four inches inside the edges of the foundation. The fences have double walls coursed on both facades and battered sides that slope inward toward the top (fig. 2.4). The fence is twenty-four to thirty-two inches wide at the base, tapering to eighteen to twenty inches at the cap, or top, course. The battering employs gravity to pull and hold the two faces together. The height from the ground to the top of the coping, or vertical topping, is about five or five and one-half feet, although there are both shorter and taller examples.

Most Inner Bluegrass rock fences are of relatively block-like quarried rock. While masons shaped the rocks to the desired size for ashlar fences, they constructed most with rock just as it arrived from the field or quarry without shaping, other than occasionally knocking off a rough corner. Each fence course is usually a uniform height the full length of a fence section, but individual courses may vary in height (fig. 2.5).[4] They range from one or two to six or eight inches tall, depending upon the natural height of the parent stratum in the quarry or the size of the rock taken from the field (figs. 2.6 and 2.7). Occasional fences have the tallest courses at the base, with the courses diminishing in height from base to cap (fig. 2.8).

Since each course forms the base for the one above, each rock must rest firmly on at least three points in the course below so that it cannot teeter.

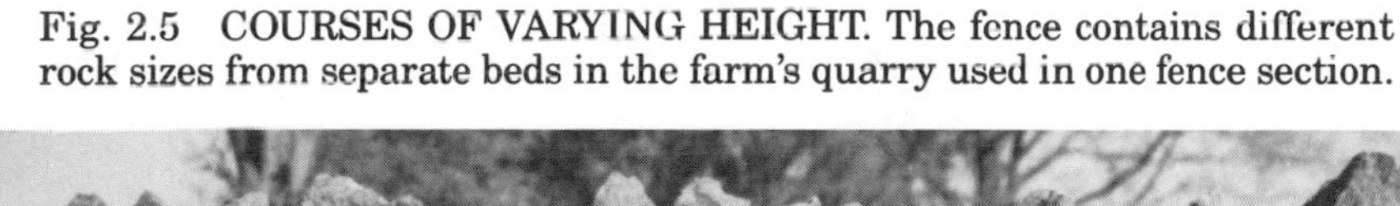

Fig. 2.5 COURSES OF VARYING HEIGHT. The fence contains different rock sizes from separate beds in the farm's quarry used in one fence section.

Fig. 2.6 LARGE COURSES

Fig. 2.7 THIN COURSES. The fence includes a mixture of waterworn, creek-bed, and field rock. *Photograph by K. Medford Moreland.*

Fig. 2.8 LARGE BASE COURSES. The mason put the largest rocks at the bottom of the fence, using smaller ones as he worked upward.

The most solidly built fences have all the rocks in each course placed with their longest dimensions running into the center of the fence; these may touch or overlap the rocks of the opposite face, which helps to bond the two sides together (fig. 2.9). A correctly bedded rock lies in the fence horizontally with its grain running in the same direction that it did in the quarry. A rock situated on its end vertically in the fence, called a "shiner," is liable to flake, or "spall" (Giles 1988) (fig. 2.10). Masons also avoid "running," or vertically aligned, joints because they form a weak perpendicular seam. Instead, they attempt to "break joint," laying one rock over two, with the one rock resting on half of each of the two in the course below it to form a covered joint (fig. 2.11).

Tie-rocks are the most important solidifying feature of a fence because they connect the two faces to each other. These are large rocks placed with their longest dimensions running entirely through the fence in a manner similar to headers in a brick wall. High-quality fences have two courses that contain ties, one about fifteen inches from the ground and another at thirty inches. The ties in each course are spaced four to six feet apart in an alternating manner, so that the tie-rocks in the upper row are in the middle

Fig. 2.9 FENCE INTERIOR. The farmer purposely made a break in this fence so that he could move equipment from one field to another. The break exposes the fence's interior, showing how the rocks meet or overlap those from the opposite face.

Fig. 2.10 "SHINER." The rock in the center of the photo has its grain running upright instead of horizontally. "Shiners" usually fill an awkward space in the coursing as this one does.

Fig. 2.11 COVERED JOINTS. An exceptionally well-built quarried rock fence in Bourbon County has every joint covered by a rock in the course above.

Fig. 2.12 TIE-ROCK SPACING. One of the best built extant fences in Kentucky, this quarried stone plantation fence in Bourbon County contains two rows of protruding tie-rocks. The ties do not protrude on the other side of the fence, which faces an abandoned road bed. The fence dates to the second quarter of the nineteenth century and has never needed repair.

Fig. 2.13 TIE-ROCK COURSE. Several farms in southwestern Clark County have fences with solid rows of tie-rocks approximately halfway up the fence.

of the distance between the tie-rocks in the lower row (fig. 2.12). Rare fences have a full course about halfway up the fence that is entirely composed of tie-rocks (fig. 2.13). On many fences, particularly the oldest ones, the masons did not square the tie-rocks off evenly with the fence face but left them protruding a few inches on one or both sides of the fence. They did this to avoid the alternative of reducing them to the width of the fence with a stone hammer, which could crack or weaken them. Protruding tie-rocks could also have served to remind the mason of the spacing as he worked and showed the owner that they were installed. Modern masons speculate that another reason for protruding ties might have been to discourage cattle from rubbing the fence. Where ties protrude on only one side of the fence, it is usually on the inside of an enclosure. Ties in road frontage fences, for instance, protrude on the side away from the road, the side not seen by passersby.

Except for protruding tie-rocks, the two outside fence faces have fairly flat planes. In contrast, the varied rock lengths inside the fence create an irregular cavity. The mason tightly packed this cavity with small rocks, known as "chinking" or "spalls," so that the center of the fence would be as solid as the faces. One contemporary farmer who repairs his own field-rock fences explained that the insides of these fences are so solid that they appear to be built from the inside out, with every rock, not just face rocks, tied to another (T. Soper 1988).

In most fences, each rock lies flat, both lengthwise and crosswise. It is difficult to detect any possible purposeful cant in the original bedding because most fences have settled over the years, altering the bedding plane, and because well-built fences have no breaks by which to inspect the interior. In a few fences it appears, however, that the individual rocks in each course cant slightly downward on the inside of the fence. By this arrangement rainwater cannot pool on the tops of the courses but drains downward through the fence and out the crevices at the bottom. The advantage of canting the courses downward on the inside of the fence rather than on the outside is that the force of gravity pulls each rock inward, another technique that adds cohesion to the fence. Most working masons today, however, advocate level bedding.

Fence courses are usually on a true lengthwise horizontal, regardless of land contour (fig. 2.14). Level lengthwise bedding also works with gravity because laying the coursing parallel to a slope's contour would direct the weight of the entire structure downhill.

The cap course and the coping also function as ties. Like tie-rocks, rocks of the cap course cover the fence's full width. Some fences have projecting cap courses that extend two to six inches beyond the fence faces (fig. 2.15). Any irregularities in coursing height are corrected by adding shallow rocks below the cap course level. This allows the cap rocks, selected for their uniform height, to form a solid base for the coping.

Fig. 2.14 COURSING ON HILLSIDE. The land descends steeply to Plum Creek Valley in Bourbon County, yet the courses are, as is usual, horizontal, not parallel to ground slope. The material is field rock. The coping slants downhill, as is typical in Kentucky.

Fig. 2.15 PROJECTING CAP COURSE. Many mid-nineteenth century fences throughout the Bluegrass have projecting cap courses; this Fayette County example extends more than most.

Fig. 2.16 TRIANGULAR COPING ROCKS. Although similar in most respects to other fences with full-width sharp-topped coping rocks, this coping on a fence in Boyle County is several inches wider than the cap course, forming an uncommon projecting coping. The multi-purpose field barn is used for curing tobacco in the fall and for livestock shelter during winter months.

The coping consists of large triangular rocks that were set aside during construction for this purpose. Spanning the full width of the fence, coping rocks are situated on their edges with the longest side of the triangle at the bottom, resting on the cap course. They lean at a slight angle, one against the other, with their tops slanting downhill. This angle beds them firmly into the cap course; if they leaned uphill, the bottoms would have a tendency to slide out, thereby upending themselves. This method also facilitated construction since the mason would begin laying the coping at the fence's downhill end at a post or corner and work uphill, leaning each rock directly against the one below.

The points created by the shorter sides of the triangular coping rocks protrude upward, forming a jagged, sharp fence top intended to deter stock (fig. 2.16). In fences with copings one large rock in width, the coping rocks serve as additional ties. This feature does not always occur on fences with a full-width cap course; many fences, especially more recent ones, have either a full-width cap course or a full-width coping. The most solidly built fences, however, have both. Uncommon alternatives to upright copings are those of large ashlar blocks laid flat on top of the cap course, spanning the full fence width (fig. 2.17).

Fig. 2.17 FLAT COPING. Most extant examples of atypical flat ashlar copings are on fences surrounding mills, churches, and graveyards. This one is at Bethel Presbyterian Church in Fayette County.

The only tool essential for building a rock fence is a rock hammer. While one contemporary mason stressed "*only* a hammer" (Kelly 1989b), a crowbar was also useful (*American Agriculturist* 1859, 110; Worford 1988). Some masons also used a batter frame, cord, and measure for sizing and perhaps a level and plumb bob (Gormley 1987). Masons built their frames of boards one inch thick and three inches wide to the size of the fence cross section. This provided a template for the fence's angles and dimensions. Historically, the mason set up the frame between the completed fence and the end of the section under construction and stretched cord between them as a guide (*Valley Farmer* 1857, 42-43). He moved the guidelines higher as the work proceeded upward. Today, essential tools are a three-pound hammer, a level, tape, and line fastened to a bar that replaces the wooden frame (Brown 1990; Guy 1990) (fig. 2.18).

In 1859 the *American Agriculturist,* having previously published a series of articles on Kentucky agriculture, described rock fence building in detail and added information about the construction process: "To make the best wall, the two sides are to be simultaneously built, with a line on each side to work by, and if two good wall builders can work together, one on each side, the same length of wall will be better and cheaper built than if but one work alone. . . . As the work goes on, two, three or four common laborers can be profitably employed in heaving over and lifting the stones from the adjoining heap to the builders by which their more valuable time may be devoted altogether to laying them in on the wall" (110).

By the mid-nineteenth century, national publications recognized the exceptional quality of Kentucky's fences. The *Valley Farmer,* a widely distributed St. Louis journal, gave them accolades and described their construction in great detail, stressing the importance of "binders" (tie-rocks) in a strong fence (1857, 42).

Turnpike Fences. Rock fence building reached its peak in Kentucky during the second half of the nineteenth century in conjunction with turnpike construction. Hundreds of stonemasons worked for turnpike companies building retaining walls to support fills and cuts that the new roadbeds required. Landowners whose property bordered these new roadways embellished their plantations by hiring masons to build road frontage fences. Rock fences became so popular that they once lined both sides of almost every lane and pike in the Bluegrass region, a fact that older residents and masons can recall (Cassidy 1989; Letton 1989; Waugh 1988).

Fig. 2.18 TOOLS. Tools pictured are those used by a modern farmer in repair of his fieldstone fences; *back row, left to right:* twelve-pound sledge hammer, rammer (to tamp soil), three-pound rock hammer (with inches marked for convenience), seven-pound sledge hammer, four-pound rock hammer, rammer, masonry hammer; *in front:* two knapping hammers.

Fig. 2.19 SPALL-FILLED COPING. This fence, built in the 1930s in Woodford County, has a coping formed of upright rocks on the fence's outside edges. The valley between the two uprights is filled with spalls.

Subtle structural differences between the turnpike fences and earlier plantation fences at nineteenth-century field edges and boundaries exist that, while imperceptible to the casual observer, are important. These modifications occurred gradually, and many fences are a blend of the two types. Turnpike fences employ some of the same features as plantation fences: double-walls, battered sides, and upright coping. But subtle changes in construction principles occurred, most of them for practical reasons.

Road frontage fences commonly have the longest length of each rock placed lengthwise along the fence face instead of projecting into the interior cavity. Large tie-rocks occur only intermittently because the mason relied on coping weight for fence stability. This practice allows more square footage to be covered with less rock. Conservation of construction material is expedient in a region where it must be obtained by quarrying, and masons adopted the practice when fence strength was no longer a prime consideration.

Masons employed helpers who poured buckets of chips, or spalls, that accumulated when face rocks were shaped, into the fence's interior cavity. This eliminated the time-consuming process of filling this space with firmly packed small rocks. George A. Martin, who wrote a manual for the construction of *Fences, Gates, and Bridges,* criticized these practices: "The mistake is sometimes made of placing all the larger stones on the outside of the wall, filling the center with small ones. . . . The foundation stones should be the largest; smaller stones packed between them are necessary to firmness" ([1887] 1974, 19).

In addition to these changes, turnpike fences seldom have a full-width cap course, and the coping, which was later called the "cap," rests directly on the top course. To reduce the need for large rocks to finish the top of the fence, the mason fabricated a coping by placing rows of rocks on top of both outer faces with their straight edges upright and filling the valley between the two uprights with spalls (fig. 2.19). When combined with the practice of placing rocks lengthwise in the face, this practice resulted in fences that have few effective ties.

When the face rocks in a course were not all the same height, the mason inserted thin rocks during lay-up to level the bed for the next course (fig. 2.20). For a finished look, it became common practice to tap spalls into any sizable interstices on the fence face (fig. 2.21). This "chinking," if driven tightly enough not to fall out, has a tendency to unseat the course rocks unless it is carefully done. Durability of turnpike fences depends on a solid foundation, which they usually have; for those built in the twentieth century, durability depends on concrete filling (fig. 2.22).

Turnpike fences are structurally different from plantation ones for several reasons. Since quarried stone went further when placed lengthwise, a given length of fence required less rock, thus reducing quarrying costs. Pouring spalls into the center of the fence was both easier and quicker than seating each piece individually; this task could be assigned to unskilled helpers, saving both time and labor. Because landowners paid most fence masons by the rod of completed work, this saving meant more compensation

Fig. 2.20 LEVELING ROCKS. Thin rock slabs in the fence bed level the tops for the course above. The opening in the fence base is a rainwater drain.

Fig. 2.21 CHINKING. The mason's helpers hammered spalls, or chinking, into the spaces between the rocks. Chinking is inserted after lay-up; leveling rocks (fig. 2.20) are placed during lay-up.

for the mason. Fence contractors would naturally wish to erect fences in an expedient manner to maximize their profits. Alternatively, if the mason were paid by the hour or day, a job completed in the shortest time meant the lowest cost to the landowner. Reducing quarrying time, pouring spalls into the fence, and eliminating seated filling saved expense, whichever method of calculating costs was used. All Kentucky masons did not simultaneously adopt these time- and labor-saving techniques, but in dated fences, the later ones more often exhibit these features (fig. 2.23).

These characteristics did not appear in all turnpike fences. Although the typical fence form changed gradually over time, some new techniques produced fences that are very sound. Fortunately, records exist to inform us of the structural specifications of a well-built turnpike fence. In building a fence without battered sides, according to a descendent of the turnpike and fence building firm of Woods and Cain, the builder dug a trench thirty inches wide, seven to eighteen inches deep, to the depth of hard soil or rock. He filled this trench with small stones or gravel and topped it with large foundation stones four to six inches in height that extended a few inches on each side beyond the fence base. The gravel bed for drainage below the foundation was ideal but not usual. The fence was the same thickness from the base to the top and built of uniformly sized rock throughout the fence; no differentiation was made between face rock and internal rock. "Upended rocks," four across, formed a coping ten to twelve inches in height (Miller 1989c).

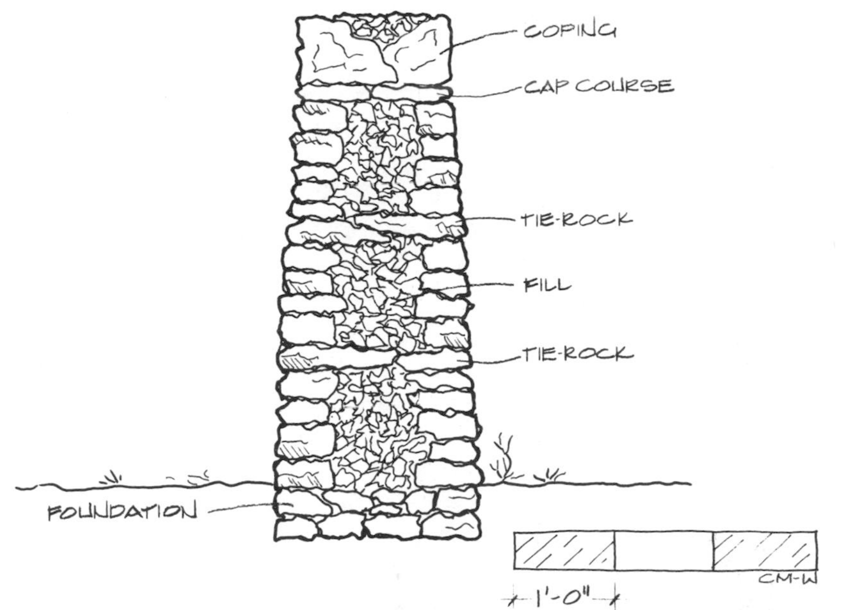

Fig. 2.22 SECTIONAL DRAWING OF A TURNPIKE FENCE

Fig. 2.23 COMPARISON OF PLANTATION AND TURNPIKE FENCES. The late eighteenth-century plantation fence on the right surrounds a graveyard; adjoining it on the left is a turnpike fence built in 1867. Differences in construction technique are evident.

Present-day farm owners sometimes request that masons use mortar in repairing old dry-laid fences. Experienced masons advise against this. New mortared walls are often built to keep water out, but old fences were built on different principles. Masons intentionally build dry-laid fences to allow water to enter and exit the fence. When mortar is later inserted in these joints, the spaces between the rocks cannot be completely sealed. In cold weather the rock contracts; water enters the fissures between rock and mortar and, as it freezes and expands, forces the rocks out of position. The freeze-thaw cycle of a long winter can cause severe damage. Falling trees and accidental blows cause less damage to dry-laid fences than to those where mortar was added; an impact can knock an extended section of rigid fence out of alignment. Mortared fences only fifty years old may require extensive and expensive repair, while older, more flexible dry-laid ones outlast them.

Characteristics of well-built dry-laid fences, whether plantation, turnpike, or modern, include solid contact between each rock and minimum chinking. Stanley Kelly, a craftsman known for his restoration work emphasizes the importance of carefully placed filling: each rock must lay flat, not on edge, to prevent them from acting as wedges and pushing the fence apart (1989b). Tom Soper, a dedicated rock fence builder and farm owner who works entirely with field rock, places emphasis on the importance of tying every rock that can be tied. He also recommends breaking the rocks as infrequently as possible and stresses the importance of covering all joints and of solid seating and wide coping stones (1988). John Guy, who is descended from a family of well-known fence masons, stresses the importance of the foundation. He digs a one to three foot deep trench for the base of a dry-laid fence, which he fills with rock or concrete, and tapers the fence sides from three feet wide at ground level to two feet wide at the top (1990).

Edge Fences. As an alternative to the flat-coursed fences built on the gentle topography of the Inner Bluegrass, mid-nineteenth century settlers in the hilly Eden Shale subregion built vertically coursed fences to confine their livestock. This type of dry-laid fence, with its upright coursing pattern, is locally called "edge fence." In addition to diagonal coursing from the ground up, structural characteristics of this fence type include a broken coursing pattern, very large tie-rocks, a downhill angle, and battered sides (fig. 2.24). Very rarely, both flat-coursed and edge fence sections occur in the same fence stretch (fig. 2.25). All recorded Kentucky edge fences contain field or loose creek rock. Masons using field rock for fences used a sledgehammer, pick, shovel (Worford 1988), and grubbing hoe for gathering the rock (*Southern Planter* 1858), in addition to a stone hammer for breaking it.

The bottom of an edge fence rests either directly on the ground surface or in a shallow trench and has no separate foundation. To begin the fence, the mason constructed a square pier on the downhill end of the fence section,

Fig. 2.24 EDGE FENCE

Fig. 2.25 EDGE AND FLAT FENCE MIXTURE. This Washington County farm contains a fine collection of rock fencing, built in both edge and flat coursing. Most rock was quarried from the adjoining creek bed and banks. A large modern silo is in the background.

laying rock in flat courses in the same manner as a quarried rock fence (fig. 2.26). He then positioned rock slabs on their edges, leaning them against the pier at a thirty- to forty-degree angle to form semi-upright diagonal courses that are often four or five rocks high, depending on the size of the rock from the fields. For the same reasons that the coping rocks of a flat-coursed fence lean downhill, the entire height of an edge fence leans downhill so that its weight is directed perpendicularly into the hillside. The direction of slant changes at the tops and bottoms of hills, accommodating fence thrust to slope angle (fig. 2.27).

Huge slabs are tie-rocks in edge fences; in places, these are large enough to span the full height and width of the fence (fig. 2.28). Unless they are large enough to reach entirely through the fence, course rocks have their longest dimension running into the fence instead of upright for maximum strength and stability. In places where there was need to dispose of much rock, edge fences may be several rocks across and three to four feet in width.

Because the field rocks used for edge fences frequently have weather-worn convex edges, they are seldom laid in continuous diagonal courses.

Fig. 2.26 EDGE FENCE PILLAR

Fig. 2.27 EDGE FENCE SLANT. From the lowest point in a valley, this Anderson County fence ascends slopes to the left and right; the rocks lean downhill on both sides.

Fig. 2.28 EDGE FENCE SLABS. These boulders in an edge fence in Owen County are unusually large, although most edge fences contain tie slabs that span the fences' full height and width.

Fig. 2.29 WEDGING ROCKS. Pointed ends of the upper rocks in an edge fence are inserted into the crevices between the rocks below, acting as keystones that tighten the fence.

Instead, the pointed edges of the upper courses are placed into the crevices between the lower rocks (fig. 2.29). This construction makes each rock a wedge that gravity pulls into any seam, relentlessly pinching the rocks together. As with a fence built of keystones, ground settling or soil shifting during a winter's freeze simply locks the fence tighter. Edge fences have battered sides, further employing gravity to tighten the fence. Some fence builders broke field rocks in half and used the broken faces on the outside of the wall, creating a smoother surface plane. The upright courses provide no place for rain and melted snow to collect and freeze, eliminating that damage factor.

Edge fences rarely have a separate vertical coping, although a few Clark County examples do. In these, the coping rocks are wider than the fence and project slightly on both sides, thus acting as ties (fig. 2.30). Such a coping may have been added to use up rock that appeared in the pastures and fields after construction of the original fence, to add extra height on uneven ground to discourage climbing and reaching animals, or merely to provide a decorative touch.

Edge fences are curiosities in Kentucky although they are found in many locations throughout the Eden Shale hills, on steep terrain near the Kentucky River gorge, and near low-lying watercourses of the Inner and Outer Bluegrass. The same basic form occurs from Robertson and Nicholas counties in the north to Mercer and Washington counties in the south and west. Residents of the mountains of eastern Kentucky also built vertically coursed fencing, there called "rick-rack" (Waugh 1988). Edge fences have been built from the mid-nineteenth century to the present on land that was usually less expensive and less desirable than that in the Inner Bluegrass. The largest area in which edge fences were exclusively built, and the area in which they are most dense today, is in western Mercer County.

On hilly land, edge fences are favored over flat-coursed ones because they utilize most easily the kind of rocks found there. Since the wedging action of the vertical courses tightens the fences over time, most last longer than horizontally coursed fences. They also "stand up better" on steep inclines (Traynor 1989). Edge fences continue to be a logical fencing choice in places where field rock is abundant. They are expedient where soils are thin and bedrock lies just below the surface because the steel posts used for modern wire fencing cannot easily be driven into the ground to a stable depth.

Fig. 2.30 SEPARATE COPING ON EDGE FENCE. Though uncommon in other sections of the Bluegrass where edge fences are located, many edge fences in southwestern Clark County have separate copings, seemingly a local trait.

Fig. 2.31 ROAD FRONTAGE FENCE. Such fences often connect to formal rock driveway entrances. Rock posts support a pasture gate and may be topped by decorative statuary.

Mortared Fences

The second rock fence type in the Bluegrass—the mortared fence, or stone-veneered wall—is a much admired feature of the region's larger farms. Most often found bordering roadways and marking entrances as decorative landscape elements (fig. 2.31), such fences also appear scattered throughout the state as popular symbols of its horse farms. Mortared fences have been fashionable since a concrete mixture based on Portland cement became a basic construction material for masonry work in the late nineteenth century. The earliest known example is a "grouted stone fence" in 1857 at "[Henry B.] Ingalls's park" on the Harrodsburg Turnpike in Fayette County (*Cincinnati Daily Gazette* 1857).

The craft of building mortared rock walls is similar to the parallel vocation of brick masonry. Skills required for mortared fences are different than for dry-laid fences, although the same mason sometimes builds both. While mortared and dry fences are two distinctly different structures, they are discussed here in sequence because general perception of the two arts is indistinct and the end product is sometimes deliberately similar in appearance.

Mortared fences fall into two general categories, those with cores of concrete block and those with cores of concrete and rubble. Choice of facing pattern may affect the choice of material for the core. Ashlar fences often have block cores and those built to look dry-laid usually have rubble cores, although the opposite is not uncommon. The earliest mortared fences, however, are usually patterned and have rubble cores. A substantial foundation is the most structurally important feature of any mortared fence, regardless of its above-ground form. Ashlar walls are faced with quarried stone, but unlike the raw material in a dry-laid fence, this stone has square corners and straight edges; careful shaping and fitting are attributes of high-quality ashlar work. These structures are technically stone fences rather than rock fences by virtue of the material having been shaped, but in popular terminology this distinction is seldom made.

Concrete-and-Rubble-Core Fences. The double wall with rubble core is the oldest and most common mortared fence variety. For this fence subtype, the foundation may be a poured concrete footer or a trench up to two feet deep packed with rock. Some masons prefer a foundation without concrete even though the fence is to be laid with cement mortar, reasoning that the ground will shift and a flexible base can respond without damage (Guy 1990). The mason lays the two outside walls on the foundation, seating each stone into mortar spread on top of the course below. He fills the cavity between the faces as the wall goes up similarly to a dry-laid fence. Instead, however, of packing the interior cavity with rocks or filling it with spalls, he fills it with a mixture of rubble rock and concrete, forming a solid core to which the face stones bond (fig. 2.32).

Fig. 2.32 MORTARED FENCE UNDER CONSTRUCTION. Masons veneered quarried rock to a concrete core at a serpentine entranceway in Fayette County.

Fig. 2.33 TWO-TO-ONE PATTERN. Large ashlar blocks laid flat form the coping for this section of the Keeneland Racecourse fence in Fayette County.

Fig. 2.34 CRAZY WORK PATTERN in Boyle County.

Fig. 2.35 SQUARE MORTAR JOINT. The raised, square cement-mortar joints sealed this fence so tightly that it could not "breathe." Water seeped into the grain of the upright coping rocks and froze, causing the crack.

There are two schools of thought regarding rainwater and mortared fences. The intention of some masons is to prevent any water from entering the wall. They solidly fill each space and seal each joint with mortar to keep water out of the fence interior, protecting it from freeze-thaw damage. Other masons use concrete and rubble in the way filling is used in a dry-laid fence, packing it into the fence interior. The soft concrete conforms to the empty spaces and when set is harder than limestone, thus acting as filling rock and forming a sturdy fence core. With these fences, water is allowed to seep into the top and out the bottom.

Tie-rocks, as in dry-laid work, may connect the two fence faces. In some fences today, curved four-foot lengths of reinforcing bar, "re-bar," are placed in a zig-zag fashion between the courses to act as ties (B. Clark 1990).

Using cement mortar between the rocks freed masons of the necessity of seating the courses directly one upon the other as in a dry-laid fence and allowed them to create a variety of coursing patterns. "Random ashlar" has variously-sized squared stones; "crazy work" and "flag pattern" are laid like flagstone; and in the "two-to-one" pattern the height of two courses butt into one stone, called a "jumper." The two-to-one pattern is said to have originated in Russia and to have been brought to America by John Oliver Keene, who built Keeneland horse farm in Fayette County. Jack Higgins built the fence at Keeneland Race Course in the 1930s with stone from a quarry on the old Keene farm (R.R. 1956; Higgins 1988) (fig. 2.33). A great advantage of the two-to-one and random ashlar patterns are that they allow all the layers of one vertical quarry cut to be used in the fence, no matter their height, rather than requiring the quarrier to obtain all the rock of one height from a single horizontal bed (Guy 1989).

With sealed joints, masons may also place stones upright in the fence face rather than horizontally as in the quarry, since mortar joints deter spalling by preventing water seepage into the stones' raw edges. This innovation made possible the crazy work and flag patterns, in which all face rocks are set vertically (fig. 2.34).

In patterned walls the mason entirely fills the spaces between each stone with mortar, creating clearly defined vertical and horizontal joints. He finishes the joints with a variety of mortar patterns, most commonly tooling them flush with the surface of the wall, a "smear" joint, or in a slightly concave profile. Raised mortar joints having squared edges, called "ribbon" and "forked" pointing, were popular around the turn of the century on decorative fences (fig. 2.35).

Large blocks of stone placed end to end were one desirable way to finish the tops of mortared fences, as they had been in some dry-laid fences. The most distinctive copings around the turn of the twentieth century, however, are castellated, with every other stone upright (fig. 2.36). Although its original purpose was to present a coping that sheep were afraid to jump

Fig. 2.36 CASTELLATED COPING. The castellated fence section fronting Castleton Farm mansion house in Fayette County dates to the early 1900s. Other fence sections on the farm, although apparently the same age, have solid upright copings. Larger blocks in the section pictured are set on edge with cement mortar in a random ashlar pattern.

(Brooks [1977] 1989, 75), this style had more to do with architectural fashion in Kentucky than with requirements of fencing.

In mortared fences, the traditional-style copings of contiguous upright rocks customarily have vertical instead of diagonally set rocks. As in turnpike fences, masons formed the coping by setting pairs of square-ended rocks upright on the top outer edge of each wall face and placing spalls into the valley between the pair to create a flat, level top. They often cover the coping tops with cement to inhibit the intrusion of rain and snow and the theft of coping rocks (fig. 2.37).

Block-Core Fences. The other and newer mortared wall system, block-core, begins on a substantial concrete footer. The rigid foundation protects the fence's mortared joints from cracking in case the ground settles. Modern masons dig the footer trench with a backhoe to a depth of eighteen to twenty-four inches and pour concrete in the trench. The footer varies in depth from twelve to sixteen inches, depending on ground slope and fence size. Two or three rows of ⅝-inch horizontal reinforcing bars strengthen the footer, sometimes cross-hatched with short pieces of re-bar every twenty-four inches. The re-bar lengths lap about a foot and are welded together or tied with heavy-gauge wire. In some cases the bars are L-shaped with the ends turned up into the wall above in order to tie the wall to the footer. If there will be any lateral pressure on the fence, such as in a retaining wall, the mason connects vertical reinforcing bars in the wall to the bars in the footer. In these, openings in the wall base allow ground water to drain through.

Fig. 2.37 MORTARED FENCE. This early twentieth-century Fayette County fence is quarried rock facing a poured concrete core.

The footer is four to six inches wider on each side than the block core, creating a ledge upon which to set the wall face. The mason constructs a twelve- to sixteen-inch wide concrete block wall, one or two blocks wide, along the center of the footer. While some masons build one or two courses of block the full width of the footer to raise the foundation to ground level, building the core and facing on that, most commence the block core and stone facing directly on the footer two to eight inches below grade. The footer is always on a true horizontal and, where the wall ascends a hill, is built in steps. Masonry ties, placed between the concrete blocks in the wall, attach the stone face to the core. Galvanized metal ties are common, but stainless steel ties are longer-lasting (Brown 1990; Carmickle 1990; Giles 1988; Harp 1990; Witt 1990) (fig. 2.38).

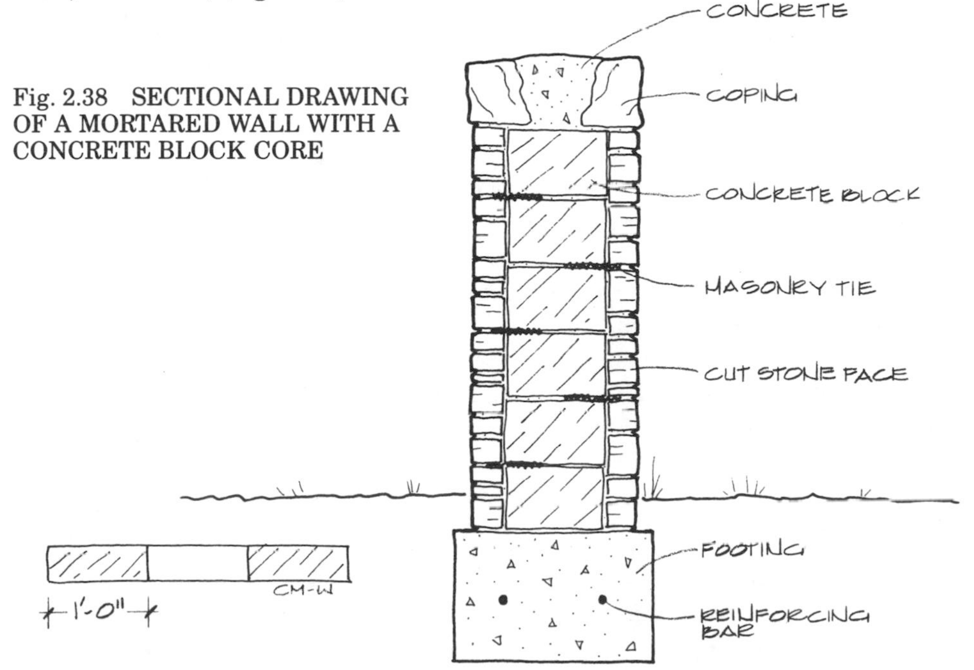

Fig. 2.38 SECTIONAL DRAWING OF A MORTARED WALL WITH A CONCRETE BLOCK CORE

Fig. 2.39 MACHINE-CUT STONE. All six sides of each stone are power sawn. Masons chipped the edges by hand to produce a rusticated fence.

Stone, veneered to the block, is four to six inches thick on the fence face, with flat surfaces on front and back. In ashlar work, each stone also has straight ends and square corners. Some masons do all stone shaping by hand. On other jobs, owners purchase fence material from a local stone company that has run it through a rock breaker to produce blocks four inches thick.

Desirable rock types for mortared fences are Oregon, Tyrone, and the Lexington Limestones, of which "bluestone" is a popular member. Oregon and Tyrone are special order items, and the stone company quarries them for each job. A supplier who gathers or quarries rock from local creeks and creek banks furnishes bluestone to the stone company. In cases where masons provide the material, they extract rock from ledge quarries, obtain it from construction projects, or purchase it from farmers.

One masonry contractor recommends a three-to-one mortar mix, three parts sand to one part Portland cement. He uses a richer mixture with less sand for pointing joints, which he finishes in a flat, raised, recessed, concave, or grapevine joint (Harp 1985, Harp 1990). Another contractor uses a proportion of twenty-five shovels of sand to a ninety-pound bag of Portland cement for the mortar mix (Brown 1990). Other masons prefer to keep the proportion and ingredients in the mix a trade secret and use a different mixture for the footer, cavity, and pointing (Guy 1990). Basic tools for shaping stone include a mash hammer, pitching tool, wedge, and tooth ax (Higgins 1988); for ashlar work, masons also use chisels, points, and bush hammers (Thierman 1959). Specifically shaped trowels provide the profile for patterned joints although some masons fabricate their own tools: a four-pronged table fork with the inner two tines removed can serve as a tool for shaping a raised, square mortar joint (Kelly 1989a). For flat or recessed joints, the fence face and seams are cleaned with a wire brush.

Any stone pattern can be veneered to the block core. "Random ashlar" is one of the most popular patterns. Rough-faced stonework is another mortared wall variation, in which the individual stones have machine-cut sides and a rusticated face (fig. 2.39). Some masons who use machine-cut stones carefully knock off the edges of each stone to create hand-tooled blocks (Witt 1990).

High-quality workmanship in ashlar walls exhibits a pleasing composition, arrangement, or "rhythm," an aesthetic difficult to define but easily recognized by experienced masons. Covered joints are important. Careful shaping to produce square corners on uncut face stones is an attribute that takes time and skill and that, although costly, is the mark of a good craftsman (Guy 1989; Harp 1990). Since it is important to block-core wall longevity to keep water out, masons take care to seal all joints and prevent intrusion of water into the structure. Good maintenance of these walls includes systematic re-pointing and removal of close growing trees and brush whose roots can dislodge the foundation.

Some attributes of well-built mortared walls are similar to those of dry-laid fences. One mason believes that use of large stones and a significant amount of contact between each denote high quality (E. Taylor 1988). Another contractor equates good workmanship with a minimum of chinking; he considers a large amount of chinking indication that the mason did not fit the rocks well (Harp 1988).

Dry-Laid-Look Fences. In contrast to patterned walls, mortared walls faced to simulate old dry-laid fences are increasingly popular. These twentieth-century fences have grown in favor to the point that they have become locally traditional. In building fences to look dry-laid, the courses are random or are laid in level rows of varying height and are veneered to a block or concrete and rubble core (fig. 2.40). The mason rakes back the mortar joints between the courses to make them inconspicuous, creating the appearance of an unmortared fence. These are "scratch" or "brushed" joints (fig. 2.41). Contemporary masons who say that their fences are dry-laid usually mean that the mortar does not show; other masons continue to build fences without mortar (fig. 2.42).

Fig. 2.40 DRY-LAID LOOK. A Woodford County fence, built in 1985 of Lexington bluestone, is veneered to a twelve-inch concrete block core.

Fig. 2.41 "BRUSH JOINT." Joints in this modern fence with the dry-laid look are recessed, making the mortar scarcely visible. It can be seen in the second row, far left, and in the bottom row to the left of center.

Fig. 2.42 NEW DRY-LAID FENCE. Built to match existing mortared sections, the new fence contains no mortar yet exhibits features typical of mortared fences: rocks running lengthwise along the fence and a spall-filled upright coping. The fence was built in 1989 in Fayette County.

When the core of a fence intended to look dry-laid is a mixture of rubble and concrete, the mason deposits it between the faces as he builds. Since these fences have shallow and deeply recessed joints, it is difficult to prevent water from entering the seams; some masons pack the mortar loosely with the intention of allowing what water enters the fence to drain from it (Creech 1990). In these, as in other mortared fences, a substantial foundation is essential.

Field rock and ledge rock are the best facing material to match the look of a dry-laid fence. Creek rock, avoided in dry work, is also suitable in mortared walls. Some mortared walls built to match plantation and turnpike fences may in fact be veneered with reused rock from dismantled fences, a practice that has obliterated many historic fences and threatens the survival of the remainder.

Special Fence Features

Several design details might be incorporated into a fence, depending upon topography and anticipated use. Special features relating to water provided for its protection and use, as well as for ways to deal with flow and runoff. To slow runoff, and consequent soil erosion from fields and pastures, farmers constructed rock dams across drains (fig. 2.43). Where a fence crossed a small drain or spring outlet, a water passage was necessary to allow the water to flow under the fence without putting pressure on it. The opening

Fig. 2.43 DRAIN DAM. A coursed rock wall, built across a drain in a Clark County pasture, slows rainwater runoff and traps topsoil. It differs from a ravine dump in which surplus rocks are simply deposited where they impede runoff, although both are placed with the same intent. Earth has built up behind this dam to allow its use as a loading ramp; truck bumpers back up to the timbers on the top of the dam.

might be nothing more than large rocks placed vertically on each side of an opening, bridged by a long horizontal stone lintel, with normal coursing above and on either side (fig. 2.44).

Larger creeks required a water gap. For these, the farmer or mason built a large stone post or abutment on each side of the creek and hung a gate from a pole or a cable stretched between the posts. Wide watercourses have additional rock posts in the creek, each a gate's width apart. The bottom of the gates rest on the surface of the water and float freely if rain brings high water. This arrangement allows water to flow without damage to the fence but prevents the passage of livestock (fig. 2.45). A description of a water gap was written by "Rusticus" (Cassius M. Clay of Madison County) for the *Kentucky Farmer:* "The best water gaps are pillars made of stone, with stone wall wings, and round, so the logs will not lodge. A good shutter is a common horse rack, suspended by hinges or wooden hooks upon a pole resting on the pillars. When the high water subsides, you may have to clear away remaining driftwood or leaves, which may prevent the gate closing tightly. This is better than going to New Orleans for your rails, or making new ones every flood!" (1859, 122-23).

Fresh springs were the primary water source on most Bluegrass farms until the mid-twentieth century, when underground water supplies became polluted. The best way to protect a spring was enclosure with stone walls (fig.

Fig. 2.44 WATER PASSAGE. A nicely constructed Woodford County water passage allows rainwater to drain from the pasture without pressure to the fence.

Fig. 2.45 WATER GAPS. A small branch of North Elkhorn Creek traverses this Scott County barnlot. Hanging water gates span the branch and allow water to flow freely. The dark line in the background is another rock fence section surrounding the pasture.

Fig. 2.46 SPRING. Several fresh water springs on this Bourbon County farm are enclosed by low rock walls.

Fig. 2.47 ROCK-LINED POND. A spring branch was channeled to support a rock-walled stock watering pond. The purpose-built pond differs from a water-filled quarry, although both are used to water stock. Algae cover the water surface in midsummer.

2.46). The enclosure was often roofed to create a cool springhouse in which to store dairy products. Some farmers also built rock retaining walls to border spring runs and sometimes channeled spring water into a rock-lined watering pond (fig. 2.47). Hog pens might be built to straddle the spring drain, providing an acceptable place for the swine to wallow.

Other devices allowed people, animals, and equipment to pass through, over, or under the fence. Where a road or lane passes through a fence, heavy square rock posts on each side of the opening support hinges and fasteners for gates (fig. 2.48). Main entrances to large farms often display large curved walls with built-in posts supporting iron gates and lamps and stone tablets

Fig. 2.48 ROCK POSTS. Most gate and end posts are square and terminate in a flat cap.

Fig. 2.49 ANIMAL PASSAGE. The sheep gap in this Bourbon County field rock fence is temporarily closed with rocks but can be opened to permit passage of sheep from one pasture to another.

bearing the farm's name. Sheep farmers built short openings in pasture fences to permit sheep to move directly from one pasture to another. These animal passages are similar to water passages except that they are taller and wider. The farmer closes the opening with temporary rocks that can easily be taken out when the opening is needed and replaced when it is not (fig. 2.49).

Fig. 2.50 STILE. The stile in this well-built quarried rock fence in Bourbon County is typical in design.

Stiles facilitate pedestrian crossing over the fence without damage to the coping. To build these, the fence mason placed long, oversized projecting tie-rocks through the fence, spacing them to form natural steps. The steps extending a foot or more through the fence on one side are the mirror image of those on the other. The mason made a gap in the upright coping at the top of the steps and placed a thick rock on the fence top at this break to form the top tread (fig. 2.50).

When building a wall along a tree line, contemporary masons build arches where the fence base abuts large trees to give the roots space to grow without damaging the fence. Where a fence ends, its face pattern continues around the end of the fence to make a "wall head" (fig. 2.51).

In very rare cases, a fence might contain a date stone, but these are difficult to find. Masons found a stone reading "J. Kearney 1863" (fig. 2.52) inside the fence at Stonewall Farm in Woodford County when they were repairing the fence and repositioned it so that the date would show (Wells 1988). Another fence in Woodford County is inscribed "J.W. McIntosh 1842" (Brown 1990). A stone in a rebuilt fence on the Leestown Pike in Fayette County shows "J. Guy 1953" as the mason (Willmott 1989).

Fig. 2.51 WALL HEAD. Where this fence section ended, probably at a gate no longer in place, coursing continued around the fence end near a stock barn. Foundation rocks, wider than the fence base, and the fence's batter profile are evident.

Fig. 2.52 *below* DATE STONE. "J. Kearney. 1863." Tool marks are also visible on this date stone.

Fence Masonry Trade

How long did it take to build a rock fence? How many workers did it require? How were masons hired? How much did it cost? Some of these questions are specifically answered in the case study of the Brutus J. Clay farm presented in Chapter 5. Mason-owner agreements from other sources also provide information about both the fence-building trade and the values that masons and clients attached to specific construction attributes. Further information comes from newspaper reports and other published accounts as well as from interviews with masons and landowners.

Time and Labor. The general consensus regarding the amount of time required to build a dry-laid fence is that an experienced man with the material at hand or a man and a helper could build a rod (sixteen and a half feet) or more per day (Garner 1984, 4; Kelly 1987; Manners 1974, 102; Rainsford-Hannay 1957, 26). In 1976 a stonemason in Scott County estimated that it would take about three weeks to build one hundred feet of dry-laid fence, which averages seven feet per day for one man (Stallons 1976). Another mason, working alone, rebuilt ninety-six cubic feet in ten hours (Brown 1990).

Mortared work takes longer. In 1988 two masons and two helpers contracted to rebuild a Bourbon County rock fence that was six-tenths of a mile long and estimated that the work would take about eight months. At an average of twenty working days per month, this amounts to about twenty feet of fence per day for four workers or five linear feet per man (Miles 1989).

Historically, the ideal working situation was for two masons to work together, one on each side of the fence, with two laborers supplying them with rock. Today it is more common for two men to do all of the work, with the mason as the lay-up man and the helper bringing rock to him as needed and doing the chinking (Higgins 1988). Fence work cannot be done on rainy days when the rocks are wet or in the winter months when the weather conditions are injurious to the mason's hands (Waugh 1988). Freezing temperatures alter the properties of the rock as well as hinder the mason (Hart 1980, 75).

It is difficult to estimate the number of fences that may have been under construction at one time in one county during the second half of the 1800s, the peak rock fence building period. Although itinerant masons were undercounted by as much as half in the censuses, forty-eight stonemasons are listed in the 1860 census of Scott County alone. If two masons worked on each fence site, twenty-four rock fences would have been under construction in the county at the time the census was taken. There could potentially have been many times that number, since the majority of stonemasons were not counted; and there could have been fewer if most stonemasons listed were not working all of the time. Such speculation is inconclusive; but the 1,258 names of stonemasons in Appendix 1 signify the immense quantity of rock work they did in the Bluegrass region.[5]

Cost. Fence-building costs and methods of calculation vary widely over time and place. In the Inner Bluegrass, prices ranged from $1.00 per day in 1818, to $.50 per yard in 1838, to $.37 per cubic yard and $2.50 to $3.00 per rod during the 1840s and 1850s (Cobbett 1828, 295; *Franklin Farmer* 1838, 2: 53; Clark 1989; Shaker Ledger Books 1839; 1847; *Valley Farmer* 1857, 42; *American Agriculturist* 1859, 110). Since some records indicate that the price was halved if the client furnished the rock and (sometimes) board for the masons, these prices imply that the fence masons customarily provided the rock.

Prices for dry-laid work increased in the twentieth century. Masons who built the fences around Xalapa farm in Bourbon County in the early 1900s earned $.25 and their helpers $.15 per hour (Waugh 1988). The price was $2.50 per rod in the 1920s in Anderson County, not including the rock. The figure varied, however, from one county to the next. A few years before, fences of similar size and complexity cost $1 per rod in Anderson County and $4 to $5 per rod in Franklin County (Worford 1988). During and after the depression of the 1930s the cost of labor was low and building in rock was not much more expensive than building in wood (Gormley 1988).

By the 1950s, rock fences in Boyle County cost $16.00 per rod (Caldwell 1989) and about half again as much in Fayette County, where the cost was $1.51 per linear foot (R.R. 1956). At the same time, masons paid by the hour earned $3.50 to $4.00 per hour in Bourbon County (Thierman 1959). Rock or stone purchased for the job was measured by the perch, 24¾ (or 25) cubic feet; and in Fayette County cost $4.00 per perch delivered to the building site (R.R. 1956). Some contemporary masons charge by the perch to quarry and lay rock (Kelly 1989b). A new fence today is $250 per rod and a rebuilt one $300 in Bourbon County (E. Conner 1989). Masons in Mercer County charge $20 to $25 per running foot for a dry-laid fence four feet tall and two feet wide at the base, tapering to fourteen to sixteen inches (Carmickle 1990).

Masons using mortar calculate prices for fence building in various ways. In 1988 fence masonry cost $10-11 per linear foot in Bourbon and Anderson counties (Miles 1989; Cutsinger 1989), excluding the cost of the materials. Contemporary masons charge $2-4 per square foot on each face in Scott and Woodford Counties respectively, and $9 in Fayette (Waugh 1988; Brown 1990; E. Taylor 1988). Labor for fences with the dry-laid look is priced at $3-3.50 per cubic foot (Brown 1990). When fence work is priced by the man-hour, most masons charge $10-11 per man, although helpers generally earn between $3 and $8 depending on ability (Miles 1989; Brown 1989; E. Conner 1989). Some fencers price by the job. Repair and rebuilding the six-tenths of a mile long fence stretch in Bourbon County, for instance, was $35,000 (Miles 1989). Another fence built by contract in Fayette County in 1988 averaged $208 per running foot plus the cost of the rock (source wishes to remain anonymous).

Lexington Cut Stone Company estimates that a ton of rock broken to four-inch thickness will cover thirty-four square feet of fence face. One ton costs $160, and cement and sand are additional materials expense (Brown 1990).

Contracts. Variations in the way masons calculate costs illustrate the variety of methods for contracting jobs—by the hour, by the job, or by the measurement of foot, rod, or perch—but most work was, and still is, done by written agreement of some sort (*Valley Farmer* 1857, 42; Waugh 1988). Several such contracts survive to provide information about both the fence-building trade and the values that masons and clients attached to specific construction attributes. The detailed contracts of 1844 and 1848 between Brutus J. Clay and Francis Thornton vary in price from $1.20 to $3.00 per rod according to who quarried the rock, paid for powder, hauled the rock, kept the tools in order, furnished the meals, and even whether Clay provided food and pasturage for the mason's horse (Clay and Thornton 1844, 1848).

Like most mid-nineteenth-century fence contracts, an 1854 contract between the Concord Christian Church in Nicholas County and H.W. Roberts specified the dimensions for a fence to be built around the graveyard: "Depth in the ground six inches and three feet wide first layer base of wall two feet, width of the top of the wall sixteen inches, heighth above the surface of the ground four feet six inches, capped with a stone from four to six inches thick and wide enough to cover the wall, and two iron gates about five feet wide" (Metzger 1989). This fence still stands. The road frontage section exhibits a coping of full-width ashlar stones laid flat. Such specificity about a fence's dimensions and structure is frequent in contracts and suggests a knowledge of what constitutes a sound fence among those hiring masons. In addition, this contract provides insight into provisions for dwelling accommodations and subsistence for temporarily employed fence masons: the fence price and method of payment to the contractor was a "subscription of $400.00 and the use of the farm for one year at $100.00" (Metzger 1989).

It was not unusual for an employer to provide living quarters for itinerant fence masons during the time they worked on the farm. The widow Wasson of Bourbon County, for example, contracted with two brothers just after the Civil War to build a rock fence around the graveyard on her farm. She paid them $65 per year and allowed them the use of four acres containing a dwelling house and a garden. In return, the masons quarried the rock, hauled it to the site, and built the fence. They started the job in 1865 and completed it four years later (W.W. Smith 1989).

Rock Fences on Four Kentucky Farms

Four farms where owners have retained and preserved plantation rock fencing—those of Walker and William T. Buckner, Barton Soper, and Samuel Butts—illustrate the various fences typical in different subregions of the Bluegrass. The Buckner farms, situated on the gently rolling land of central Bourbon County, are distinguished examples of quarried rock fence building technique and placement in the Inner Bluegrass. Two farms from widely separated parts of the Eden Shale hills contain much different fencing types: the Barton Soper farm in southeastern Bourbon County exemplifies a nineteenth-century farm replete with excellent flat-coursed field rock fences. Edge fences, characteristic in the western section of the Eden Shale, are portrayed by the fine assembly at the farm of Samuel Butts in Anderson County.

The Walker and William T. Buckner Farms, Bourbon County. A few miles east of Paris, the county seat, between the Cane Ridge and Harrods Creek roads, are adjoining farms developed by Walker and William T. Buckner. These farms occupy land that Walker Buckner began buying in 1817: the dates attributed to the two main dwellings are 1820-1836 and 1841. The Buckner farms lie on the high-quality Maury-McAfee soil characteristic of the Inner Bluegrass where rock is found at the surface only along steep-sided stream courses. Fences on these farms are therefore of quarried stone. Numerous hillside quarries remain visible along the boundary fence between the two farms. Masons or quarrymen opened these quarries to supply rock close to the fence construction sites, and the largest one remains open today.

All of the rock fences on the Buckner farms are horizontally coursed and built of squared stones, so they are true "stone fences." The use of quarried stone enabled the masons, who probably worked during the mid-nineteenth century, to achieve exceptionally straight and tight-fitting coursing. By using quarried stone and large tie-rocks that extend completely through the fence at regular intervals, the masons built fences that retain structural integrity and remain among the best constructed fences to be found across the entire region more than a century after completion. The fences on these two farms are boundary, interior field, paddock, and yard fences (figs. 2.53 and 2.55).[6] Some original enclosures have been removed and replaced by wooden post and plank fences. Others were pushed down and buried as the owners consolidated fields into larger parcels to permit more efficient cultivation.

The Barton Soper Farm, Bourbon County. Barton Soper inherited land east of Little Rock, in the Eden Shale section of southeastern Bourbon County, from his father, Laurence, in the mid-1800s. The family expanded their holdings over the years, and in 1915 Ivan Soper bought additional acreage that had once belonged to a kinsman, Laban Letton. Today this stock farm comprises 283 acres operated by John Soper and his son Tom. The Sopers maintain one of the most extensive rock fence complexes within the Bluegrass region, 1,330 rods or about 4.16 miles (figs. 2.54 and 2.56).

Rock fence construction began here in the 1850s, when land deeds cite rock fences for the first time. U.S. census records suggest that building

Fig. 2.53 BUCKNER FARM. Ledge rock fences surround the barnyards, house yard, and pastures. The uncommon banked stock barn in the background has ground floor walls of coursed rock. Hay storage is on the upper level.

Fig. 2.54 SOPER FARM. Carefully maintained field rock fences appear as dark lines across the fields. The burley tobacco barn in the valley was moved to this site in the 1920s and erected on a coursed rock foundation.

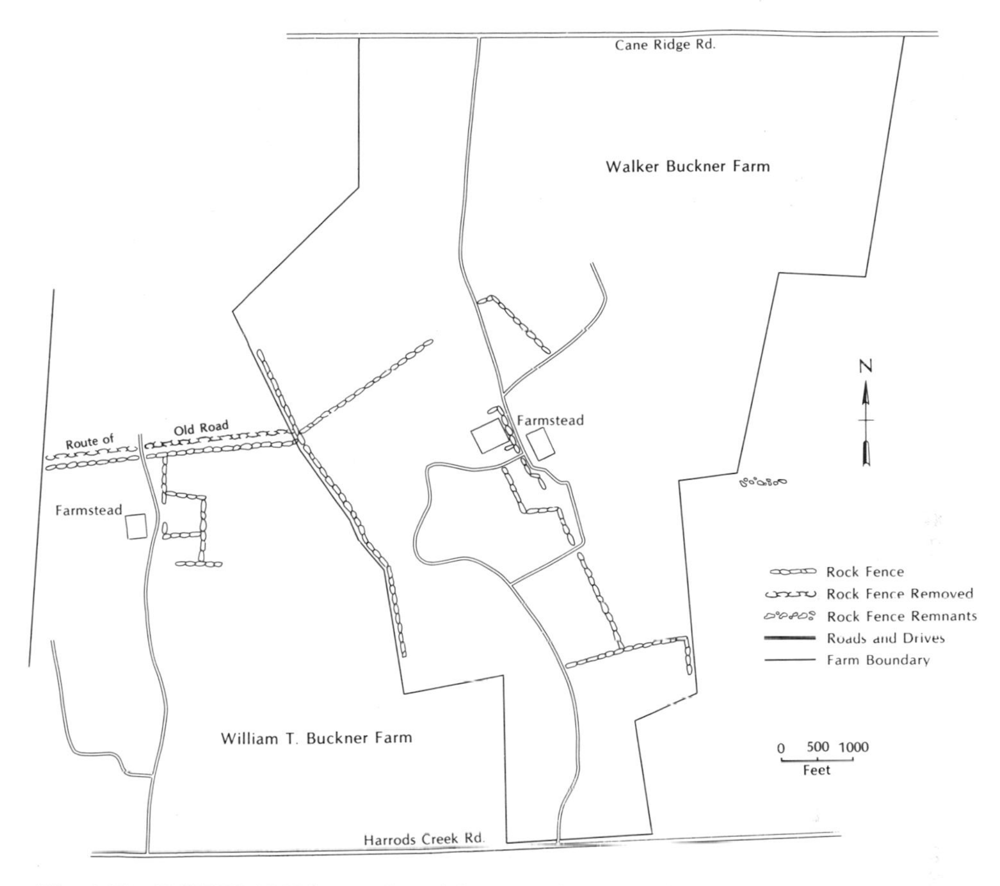

Fig. 2.55 BOUNDARY, lot, and road frontage fences on the Buckner farms as they presently exist. Note: because of land subdivision or aggregation or because of changes in road access and alignment, historic field and farmstead locations may be different from those illustrated here and in Figs. 2.56 and 2.57.

Fig. 2.56 BOUNDARY, lot, and road frontage fences on the Soper farm as they presently exist. See note under Fig. 2.55.

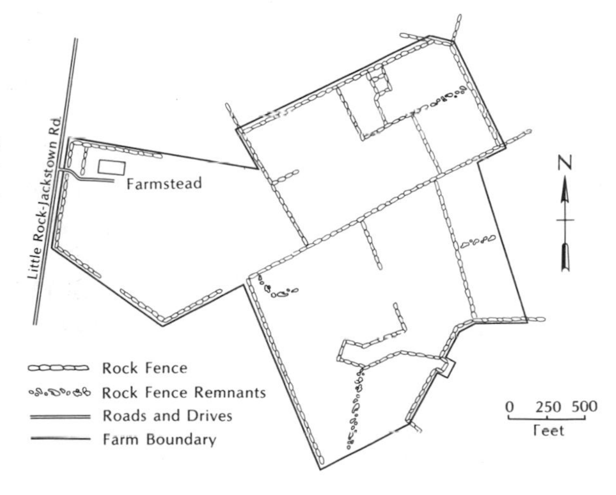

continued into the 1880s: this enumeration lists the stonemason, Richard Long, as living on the farm.

Most fences are built from field rock gathered off the steepest hillsides. The masons laid all of the farm fences in horizontal coursing, a difficult feat using irregular field rock when building on steep slopes. Fence height (an average of about forty-eight inches) and width (about eighteen inches at the top of the fence) is uniform across the farm, although fences around springs and near the creek that fronts the property are shorter and thicker.

The general pattern of odd property boundary angles, most of them bordered by rock fences, reflects the metes and bounds surveying system used throughout the Bluegrass. Some interior field fences follow ridge lines, while others run directly up and down steep slopes. Some cross creeks and turn corners. Short fences enclosing springs, spring runs, and stock lots still stand near two former house sites, the northernmost one of which is Barton Soper's. A few fences run straight for considerable distances; the line that bisects the farm includes about 3,110 feet of rock fence and continues on beyond the farm's present boundaries. This line is a segment of a larger tract that was surveyed during the initial settlement period in the late eighteenth century.

The Samuel Butts Farm, Anderson County. Samuel Butts established a farm in 1800 along the Bond's Mill-Fox Creek Pike southwest of Lawrenceburg, the Anderson County seat. He came to Kentucky from Culpeper County, Virginia, and purchased this land, which is partially in the Eden Shale hills and partially in the Salt River Valley. The hill land has thin Eden soils on heavily rolling slopes and field rock near the surface. Masons built the farm's large concentration of rock fences to line the pastures, fields, graveyard, barn lot, and property boundaries (fig. 2.57).

Neither the 1800 deed to Butts nor descriptions of additional purchases he made in 1819 when this land was still a part of Franklin County mention rock fences. Such fences do, however, comprise a portion of the property's boundaries described in the first deed after Butts's death in 1847. Deed entries thus establish this property's rock fence construction as between 1820 and 1847.

The fencing on the Butts farm is different from the horizontally coursed fences common in the Inner Bluegrass counties; most is vertically coursed edge fence. This fencing represents the functional adaptation to the Eden Shale's steep slopes (fig. 2.58). While field and yard fences have been removed, the remaining fences are in exceptionally good condition considering their age. Charles Worford repaired the farm's fences as a boy in the 1930s, having learned the trade from his father who was a fence mason in Franklin County. His northern Irish-descended family came to Kentucky in the 1700s (Worford 1988).

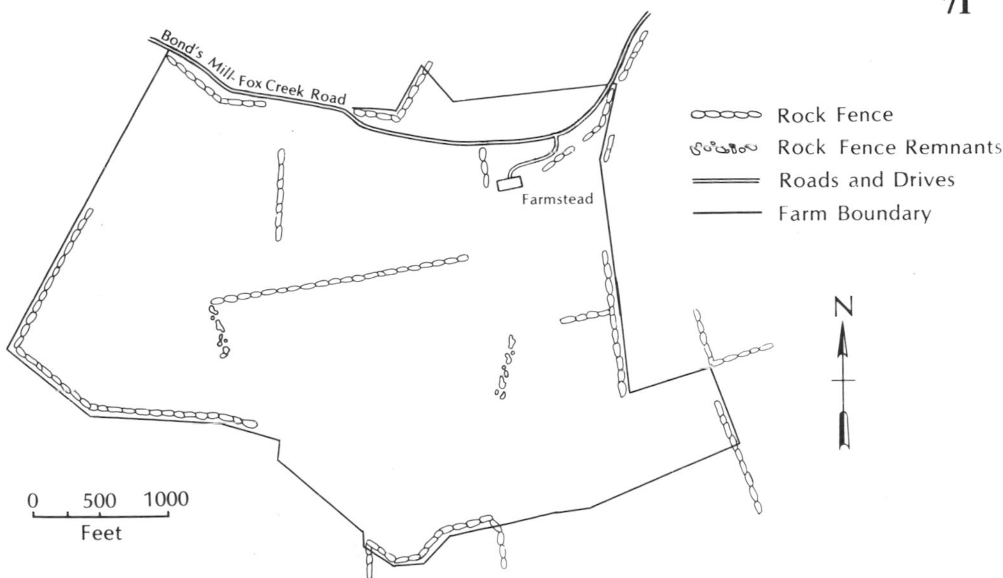

Fig. 2.57 BOUNDARY, lot, and road frontage fences on the Butts farm as they presently exist. See note under Fig. 2.55.

Fig. 2.58 BUTTS FARM. Field rock edge fence seldom requires repair and has been in use here since the mid-nineteenth century.

Soper, Butts, and the Buckners had different reasons for choosing rock and stone as fence building materials. For Soper and Butts, using rock was practical and expedient. The Buckners, in contrast, went to great lengths to obtain stone for fence building, and their high-quality fences reflect an acquired taste for "correct" appearance. But for all of them, traditions affected their choices.

CHAPTER 3

Bluegrass Fencing Traditions

> Nothing fixes our estimate of a planter sooner than the character of his enclosures. A big, high, well laid fence, with good gates and fixtures, can hide a multitude of failures at other points.
>
> —*Soil of the South,* 1854

The process of building fences to enclose land, demarcate farmyard lots, or line turnpikes was not the straightforward endeavor it may seem. It derived from the farmers' cultural traditions, the type of agricultural economy they developed, the legal code they adopted, even the values they held. Furthermore, the physical qualities of the farmland—hill slope, soil thickness, and fertility—affected choices of crops and livestock and the availability of fence-building materials. To understand why farmers built rock fences, choosing this fencing material over another, requires that we place these people within the historical and geographical context that affected their decisions and behavior.

Land Acquisition

The Bluegrass was never a place to buy land cheaply. Land acquisition was the primary influence on rapid development in central Kentucky during the last quarter of the eighteenth century, and many frontiersmen took advantage of the financial opportunities available in land dealing as well as in agriculture.

Land prices quickly rose beyond the reach of poorer settlers. The result was that many early landowners around Lexington were the sons and daughters of colonial frontier planters from the Valley of Virginia west of the Piedmont. Some early pioneers purchased Bluegrass acreage with proceeds from the sale of farms and businesses in their homelands. Others qualified to receive colonial land because they financed the importation of settlers to America, primarily from Ulster in the north of Ireland. The Breckinridge family of Ulster, who came to Kentucky from Albemarle County, Virginia, is a well-documented example (Klotter 1986). Other prominent Ulster families from west of Virginia's Blue Ridge Mountains who were influential leaders in Kentucky, in the periods both before and after statehood, include such well-known families as the McDowells (Armstrong 1878, 36), the Prestons (Brown 1870, 5), and the Browns (Hardin 1981). Many Bluegrass landowners acquired several hundred or even a few thousand acres (Jillson [1926] 1972; Brooks-Smith 1962) and referred to their properties as "plantations."

In contrast, most people of English origin who came to Kentucky did so after initial settlement and purchased already patented lands (Kentucky Historical Society; Perrin 1882a, 1882b). Virginia planters, having exhausted their land's fertility by intensive tobacco production and having divided it through the generations to sizes too small for profitable farming, looked to Kentucky's rich lands as a solution. Bluegrass fertility was renowned, and once the danger of Indian warfare was over, the area attracted migrants who were often younger sons of colonial families. Some who eventually became the most socially and economically prominent "gentlemen" farmers of Kentucky in the 1800s were descended from these families. Kentucky settlers also included families of German, Dutch, French, and Welsh ancestry, although they were a minority (Murray-Wooley 1988).

Farms and Plantations

The quality Bluegrass limestone land was especially attractive to those who could afford its high price. Speculators arrived early and brokered land to later migrants. Richard Clough Anderson, Sr., for example, who was the principal surveyor of the Virginia territory in Kentucky, established his office in Louisville in 1783. His personal speculative mania brought him to the brink of financial ruin (Tischendorf and Parks 1964). Other speculators such as the Scotsman Robert Craddock accumulated large fortunes by purchasing land warrants from Revolutionary soldiers who did not care to move to Kentucky (Cherry 1930). By 1800, prices as high as $100 per acre were common, and even wealthier planters could afford no more than a few hundred acres (Troutman 1968, 368).

The English- and Ulster-descended migrants who established plantations were often well educated, perhaps classically trained in England's and Scotland's best schools. They differed from the settlers moving with the frontier across the Upland South who practiced a skinning and exhausting agriculture that quickly depleted the soil (*American Agriculturist* 1842, 138). Rather, the Bluegrass planters initiated a farming economy based on livestock grazing and a variety of crops—wheat, rye, barley, hemp, and tobacco. Neighborhood mills ground wheat into flour for local consumption. Distillers bought rye and barley to make bourbons, which they marketed as near as the closest crossroads settlement and as far as New Orleans. A variety of Virginia tobacco was salable for cash, and hemp became a popular fiber crop that provided material for local rope and bagging manufacturers.

The profits from these products were often high and helped these farmers establish sound financial standing.

Two other livestock feed crops became the central focus of the agricultural economy: grass and maize. On Inner Bluegrass limestone soils, corn might yield over one hundred bushels per acre (*American Agriculturist* 1842, 69), and it became the preferred winter feed for cattle. The favored pasture—the wellspring of the region's reputation—was bluegrass, although other grasses and clover were also commonly planted (Henlein 1959, 4-8).

Livestock were particularly well suited to forage the rolling grasslands. Farmers raised sheep, horses, mules, and cattle, and, by 1785, imported cattle, which soon became the basis for a profitable grazing economy (Rice 1951). Not satisfied with the mediocre performance of their native stock, some farmers organized livestock improvement associations to import blooded cattle. Kentucky buyers traveled regularly to England to select prize stock (Estes 1958).

Stockmen, anticipating the demand for select stock in Kentucky and in adjoining states, adopted English-style breeding techniques and a habit of keeping meticulous records. With an economy based on English stock and a landscape that bore a strong resemblance to the pastured lands of Britain, the region's cattlemen gave the Bluegrass the sobriquet "England of America," intending to elevate the image of the Bluegrass country to equivalence with the British isle they sought to emulate.

The Importance of Fencing

A British tradition that accompanied settlers to Virginia was the practice of allowing livestock freedom to range over community pastures. Under English common law, and later American basic law, a farmer had no obligation to fence in his stock or to fence a crop against someone else's animals (Danhof 1944, 174). As long as neighbors were few and the land remained open, cattle could be branded or earmarked and left to roam at will (Long 1961, 33).

Because of the open range tradition, Kentuckians needed fencing as early as the second year of occupation. Since cattle often grazed down the corn plot or kitchen garden, farmers needed to enclose these areas from the predations of their own animals or of a neighbor's stock. Alternatively, they could fence the cattle into a pasture. At any given time, either of these tactics might be in use, depending upon the density of settlement and the time and labor available. In central Kentucky, cattle grazing became a matter of serious concern by the 1790s. Disputes between stock owners and those whose crops had been damaged were common.

In January of 1798, the state General Assembly passed an act, adopted from Virginia statutes, that defined trespass (Warder 1858, 157). The act's authors apparently recognized that a cattleman could not be blamed for his stock's wanderings if a fence were inadequate to stop them. The law required property owners to erect a substantial barrier, or they could not invoke legal protection. The law stipulated "that if any horses, mares, cattle, hogs, sheep, or goats, shall break into any grounds, being enclosed with a strong and sound fence, five feet high, and so close, that the beasts, breaking into the same, could not creep through; or with a hedge, two feet high upon a ditch three feet deep, and three feet broad,[1] or instead of such a hedge, a rail fence of two feet and a half high . . . [that fence] shall be counted a lawful fence" (*Laws of Kentucky* 1799, 27).

The law specified substantial penalties in the event a lawful fence did not adequately withstand marauding stock. Farmers whose livestock broke down such a fence and damaged crops were liable for full damages. If those same stock trespassed a second time, double damages could be collected. In a third transgression, the enclosure owner could, at his option, kill the offending livestock and then sue the owner for triple damages (*Laws of Kentucky* 1799, 27).

Strong laws and penalties evidently did not stop open grazing. In 1830 the editor of the *Western Agriculturist* criticized the practice of turning stock out to "forage or die" as a threat to farmer's crops and the "quietude among his neighbours" (Hamilton County Agricultural Society, 54). This enclosure law was amended in 1852 to clarify further the rights of the stock owner should the fence in question not meet the standards of a legal fence. If the landowner's fence did not comply with the criteria defined for height and strength, and he injured or killed the stock grazing his crops, he was liable to the stockman for double their value (Wickliffe, Turner, and Nicholas 1852, 407). Law also defined materials and techniques that could be used to build a legal fence: rails, brick, plank, a raised ditch and hedge barrier, or stone all were acceptable.

The landowner whose enclosures did not meet the standards of a legal fence put his crops in the way of wandering stock with no legal recourse. But a stockman could find himself in court with substantial expenses should his animals push through a neighbor's legal fence. Equally disconcerting, the stockman could no longer vouch for the bloodlines of cattle that broke out of their enclosure. Where purebred cattle represented a great investment, a farmer with wandering stock risked his business reputation. Good fences were needed.

Although few farmers' records have survived that allow examination of the land clearing and fencing process, travelers often remarked on the extent and quality of the enclosures they saw in the Bluegrass countryside. Alexander Wilson, a well-known Scottish traveler, visited Lexington in 1810 and observed that the fields around the village looked neat and well fenced (Coleman 1953, 267).

By 1835, Charles Hoffman found split rail or "worm fences" common along the road between Lexington and Frankfort, where he saw a hundred-acre woodland enclosed by a fence and separated from a wooded pasture where a herd of elk grazed (Hoffman 1835, 130-31). When he traveled east from Lexington, he found "enclosures, which were generally shut in by a worm fence on either side, [and which] were exceedingly beautiful" (137). Hoffman's observations are especially useful because he consistently recorded details of the countryside that others may have passed over as inconsequential. Given the level of detail he recorded, it seems unlikely that Hoffman would have ignored rock fences had they been widespread along the Inner Bluegrass roadways. Split wooden rail was the only fence material he noticed along the roads he traveled, and the rails were assembled into Virginia-style worm fences. Another traveler wrote that rail fences west of the Cumberlands were frequently six to nine rails high and were not expensive where timber was plentiful (Woods [1821] 1904, 198).

Building wooden fences was customary for Kentucky's early settlers. Most any farmer or tenant could erect one. Many farmers moving west to the Kentucky frontier were coming from a woodland tradition (Kniffen and Glassie 1966, 40). Whether from the Great Valley of Virginia, from Pennsylvania, or from New England, Kentucky's settlers were accustomed to tools and living habits that were closely tied to wood. Not only did they clear the land of trees to create open fields, but their first homes were often walled with log, while fireplaces burned wood year round. Wagons, sledges, ox yokes, hoe handles, and spinning wheels all were wood. Kentucky settlers considered axes, wedges, and saws more important than plows and shovels; most men would have owned these basic tools and had experience in woodcraft. It is not surprising that they followed familiar practices in the Bluegrass, clearing the heavily wooded sections of trees and splitting the trunks into fence rails. One cannot, however, make the same case for building with rock or stone. Although some craftsmen were skilled in stone house and rock fence construction, few farmers were masons, and building with rock was something they either had to learn or hire craftsmen to do.

From Rail to Rock

Although central Kentucky's rail fences were a useful by-product of land clearing, they had limitations. Some woods rotted quickly. A fall or winter pasture grass fire could burn rods of seasoned rail fence. When wood was in short supply, rail fences might even be stolen for fuel. A farm journalist criticized the expense of erecting and maintaining wooden fences and urged that a substitute for them be found because "[of] what ever wood they are made, however substantially they may be executed, or in whatever situation they may be placed, their decay commences the instant they are erected" (Hamilton County Agricultural Society 1830, 55). A post and rail fence might have to be renewed as many as six times a century (67). Even bad weather affected rail fences adversely. In May of 1843, a "terrible storm" near Lexington blew down fences, and on at least one farm, "not a pannel of fence anywhere [was] left standing" (Dudley 1843).

Farmers found the rail fence convenient to build but a burden to maintain. Continual rail renewal was a drain of both labor and the pocketbook. The timber supply steadily declined and wood became more expensive. By 1830, a respected farm journal noted: "Such has been the excessive and wanton destruction of the forest trees in [Kentucky], that although it is less than half a century since the silence of the wilderness was first broken by the sound of an axe, it is a lamentable fact, that many farms are already without the timber necessary for another renewal of their fences" (Hamilton County Agricultural Society 1830, 178). To purchase and haul fence timber over poor roads was an expensive alternative.

Farmers on the Bluegrass limestone lands had two options: either keep about one-third of a farm in woods to supply rail timber or replace the rails with rock fences (Beatty 1844, 260). Emphatic support for the second strategy came from an editorial in the *Southen Planter.* "He who has the stone should put them into a fence," argued the essayist, "particularly if he is scarce of timber. And if he has the timber, better sell it with the land, and expend the proceeds in stone work" (1858, 497). A writer for the *Farmer's Cabinet* observed in 1836 that replacing rail fences with stone had two virtues: the fence was durable and the construction material was simply cleared from the fields, making them easier to till and more productive (G. 1836, 57). The editor of the *Western Agriculturist* advocated building stone fences in areas where stone could be easily obtained: "They are at once a barrier, and allow the immediate use of the enclosures. They occupy but little space, are more secure than most other fences [and] are unquestionably more economical than the prevailing 'worm' and post and rail fences of the West" (Hamilton County Agricultural Society 1830, 57). An article written in the *American Agriculturist* in 1859 supported these views by saying that the stone wall was a better and cheaper fence than Virginia rails (18, 174). It was less expensive, however, only if the price of labor were low or if the expense were amortized over a number of years (Danhof 1944, 180). Further, many Bluegrass farmers had to quarry rock for fences thereby adding time and expense to a building project. This suggests that the region's farmers were strongly motivated to build rock fences. Beyond practicality, clearing rock from one's fields and retaining the timber came to be viewed almost as a moral obligation. As the *Southern Planter* put it, "For every stone which you pick up and move you shall have a credit; for every valuable tree which you cut you shall have a debt" (1858, 497).

Journal editors consistently advocated solving the wood shortage problem by building rock fences where rock and labor were available. Their views are largely pragmatic; primary concern was the practical solution of common farming problems, and they realized that in much of the Bluegrass, fence rock had to be quarried. Editors did not advocate rock fences because they provided work for a labor force—slaves in the South—during the slack winter months. Nor did journals support rock fences merely because they were handsome additions to property. The large estate owners in Kentucky's Inner Bluegrass subscribed to the nineteenth-century agricultural press and supported it by contributing information and articles. As they confronted the wood shortage problem their decision to build rock fences was simply a rational solution, supported by the agricultural press.

In the Bluegrass, farmers began replacing rail fences with rock during the early 1840s. A.B. Allen, editor of the *American Agriculturist,* praised the region's woodland pastures, the macadamized roads, the brick and stone mansion houses, and the beautiful plantations "strongly fenced in with high stone walls" (1842, 140). He recognized that the Bluegrass parks did more than merely shelter cattle and grass from summer sun, for the locust groves in the pastures "furnish timber for building and fencing stuff" (69). Allen's report may imply that while some farms had stone fences by the 1840s, rock was not considered a pervasive building material but simply came into more common use. Another interpretation is that as the popularity of rail fences declined, rock systematically replaced wood. In any case, Allen's visit placed him in the midst of the transition from rail to rock fences.

Kentucky had prototypes for rock fences from the earliest years of settlement. Scattered late eighteenth- and early nineteenth-century records mention rock fences or "stone walls" such as the stone wall at Harrodsburg in 1777 and a stone fence at Lexington in 1780 (Draper 12C:26-29; Perrin 1882b, 575). Masons built a stone wall in 1800 around the state penitentiary in Frankfort (Kramer 1986, 53) and around the jail in Paris, Bourbon County (Everman 1977, 25). David Meade surrounded his estate in Jessamine County with a rock fence before 1814 (Peet 1883, 54). By 1826, the Shaker settlement at Pleasant Hill in Mercer County had nearly forty miles of rock fences (Thomas and Thomas 1973, 15, 99).

Extant records establish a time period in which rock fence construction took place on other properties. County deed books show that the mid-nineteenth century was the primary period for rock fence construction in Kentucky. When a property was sold, surveyors customarily recorded the fencing material demarcating property boundaries, and these notes became part of the recorded deed. By comparing deeds for specific properties in which no rock fences are mentioned with later deeds that do cite them, it is possible to deduce the dates between which the rock fences on the property were built. For example, deed transfers before 1846 for eight properties near the village of Great Crossing in Scott County on which substantial lengths of rock fence presently stand do not mention these fences. References to rock fences do appear, however, in the boundary descriptions after that date, which establishes the mid-nineteenth century as the construction period on these farms (Bevins 1989). It is rarely possible to date the construction of any particular rock fence precisely from physical evidence alone because masons employed some of the same techniques over a long period, and even the same rocks have been reused in careful rebuilding and repair.

Roads and Fences

Another major catalyst for the construction of Bluegrass rock fences was masonry associated with road work. Kentucky's legislature passed into law in 1797 road building specifications that applied to the new public roads accessing courthouses, public warehouses, landings and ferries, mills, and coal and iron works (Crump 1895, 7). Roads were to be thirty feet wide with grades of no more than two degrees preferred. A road of reasonable width and shallow grade required considerable excavation. Where a road crossed a creek valley, the grade was flattened by cut and fill, and the fill sections had to be retained by rock walls. Where a bridge was needed, abutments and piers were built of stone. All of this work required stonemasons.

One consequence of these requirements was that during the early 1800s, a close link existed between the road and fence builders and between the rock used in both kinds of construction. Scots engineer John MacAdam's system of hard-surfacing roadbeds required a substantial supply of rock that had to be broken to specific size and placed in position. Stonemasons, familar with stone cleavage planes, quarry work, and the trade's tools, were logical craftsmen for road building, especially when other kinds of stone work were not available (Melish [1812] 1818, 623; Vigne 1832, 85). Some stone workers broke rock for the roadbed while others built bridge piers, abutments, and roadside retaining walls.

Workers often built road frontage fences on properties along the turnpikes concurrent with the roads. Two entries in the 1848 Fayette County Order Book, during the period when most road frontage fences were constructed, illustrate the process. In August the court appointed a committee "to examine and report to the Court, on the expediency or inexpediency of reducing the width of the old Frankfort Road to thirty feet" (239). In September, the committee reported back to the court:

> After due consideration thereof—It is ordered that the several proprietors of land on the old Frankfort Road who may desire to enclose their lands on the said road by stone fencing, may be permitted to build their stone fences so as to leave thirty feet clear between the fences for said road, provided that those who may avail themselves of this priviledge shall be required to cut down

> the banks and projections of the road to the level of the center of the road out to the said fences, and this priviledge shall only extend to those who comply with this proviso. [248]

Land owners along the Old Frankfort Pike were not alone in contracting skilled turnpike crews to build fences for the enhancement of road frontages. Farmers soon realized that road frontage fences did more than simply confine stock or provide them some shelter against cold winds. Well-built rock fences embellished the appearance of their properties and therefore came to have a symbolic value that augmented their practical use as enclosures (Mitchell 1875, 180; *American Agriculturist* 1859, 175).

The theme of fence appearance and construction quality appeared frequently in agricultural press articles regarding Bluegrass farms. An essay in the *Valley Farmer,* published in St. Louis in 1857, reported that Robert Aitcheson Alexander of Woodford County was building fences along the Old Frankfort Pike that were "constructed in the best possible manner" and that "in no country have we ever seen better stone walls than are now built in Kentucky" (42-43) (fig. 3.1). The *American Agriculturist* also had high praise for the Bluegrass rock fences: "The walls there are of quarried

Fig. 3.1 WOODBURN FARM FENCE. The quarried fence commissioned by Robert Aitcheson Alexander at his farm in Woodford County still stands. In this section, a full-width cap course supports the coping rocks.

Fig. 3.2 STONEWALL FARM FENCE. The road frontage fence at Stonewall Farm in Woodford County was completed about 1863. Its superb construction features quarried rock fractured into symmetrical blocks, placement of the thickest courses at the base, tie-rocks that protrude away from the road, and a full-width cap course. Masons also skillfully employed thin rocks to level the courses.

stone chiefly, and built about five feet high, in the most substantial manner—apparently for ages" (1859, 110).

Even the Civil War did not stop wealthy farmers from constructing rock fences. A Woodford County historian commented about Warren Viley, owner of "Stonewall Farm," that "during the Civil War the master of the estate had a stone wall built the full length of the frontage of his farm that has received more favorable comment than any masonry constructed in the county" (Railey [1938] 1968, 77) (fig. 3.2).

Fences as Symbols

During the nineteenth century, the rock fence came to be much more than a mere enclosure. It allowed passersby to compare one farmer's husbandry skills with another's (Pocius 1977, 9). Journal editors might argue that rock fences were cheaper to build and maintain over the years than rails, but Bluegrass planters who regularly read the agricultural press could see that those rock structures also addressed an emerging concern for appearance. This was considerable incentive to a reputation-conscious Kentucky stock farmer. While journals could suggest that the merit of the rock fence centered on its durability, it was also true that fence construction on a given property might take several years. This meant that fences were a substantial investment of money and management time. Well-built fences required skilled masons and therefore represented a kind of conspicuous consumption of labor. The public came to associate the rock fence with wealth, desired social status, and the region's landed plantation families. People admired the high-quality rock fence for its permanence and its picturesque qualities, and so it symbolized taste, refinement, and a concern for the aesthetic over the mundane.

Turnpike fences were a pragmatic solution to the need to enclose a stock farm's road frontage, and because of their position, these fences were the ones most frequently in public view. The rationale for building the pike fence became more than utility, it was a product of a concern for ornamentation. An irony is that as the fence came to symbolize elite values, its structual makeup was increasingly compromised. Younger fence builders had less opportunity for instruction from accomplished senior masons. Some later fences were probably built by laborers who had little training or supervision. Contractors with large labor crews hired unskilled men as helpers and apprentices in the late nineteenth and early twentieth centuries (Gormley 1988; J. Miller 1989b). More important, if landowners did not understand the subtle details of construction that produced a long-lasting fence or were simply interested in low cost and attractive appearance, they would allow established practices to be set aside in favor of speed. The purpose of many turnpike fences was an attractive road frontage rather than to confine stock (Williamson 1967); thus, strength was not the primary consideration.

After the early 1900s, the diffusion linkages that spread fence-building techniques from one location to another and from one generation to the next were largely truncated. Working masons today—with some exceptions—have no family or cultural history of rock work. Instead, today's fence-building techniques derive from the parallel vocation of brick masonry, using methods that craftsmen who began as brick masons found effective.

Rock fences filled two practical needs on nineteenth-century Bluegrass farms. First, as farm population density increased, traditional open-range grazing was no longer an acceptable practice. The laws requiring fence construction meant that farmers had to feed stock through the winter on their own acreage rather than allowing them to roam and forage. Inner Bluegrass farmers imported blooded cattle and horses for breeding and sold stock to other breeders as well as to market. Since a soundly built rock fence would contain livestock of any size, farmers with reliable fences could vouch for their stock's bloodlines, a critical consideration. Quality fences meant fewer lawsuits and reliable stock bloodlines. All this fostered stability, and stability—even power—was clearly represented by the rock fences that became the signature of the region's larger farms.

Second, the transition from rail fences to rock was also a practical solution to the increasing shortage of timber suitable for fence material in the Inner Bluegrass. Rail fences fulfilled fence law requirements, but, as land clearing continued, timber for rails was in short supply. If cheap but skilled labor could be found, each farm had a ready supply of material for building long-lasting rock fences. Labor costs were potentially a significant portion of building expense, especially if skilled craftsmen were required.

Who, then, were these fence masons?

CHAPTER 4

Origins of Fence Masons

> We cannot regard it as a matter of astonishment that so large a proportion of the emigrants from Ireland should proceed to the United States: nor can we think it probable that the stream of emigration could now be turned aside.
>
> —John S. Cummins, 1860

The idea that Kentucky's rock fences are "slave fences" has arisen relatively recently and is often repeated to tourists and newcomers. Historian Thomas D. Clark, author Samuel B. Cassidy, and restorationist Stanley Kelly, among many others, however, contend that African-American slaves did not build the fences. Older masons who know the origin of their skills and elderly residents knowledgeable about local history maintain that most of the existing nineteenth-century fences were built by Irishmen (Cassidy 1989; Hockensmith 1989; Kelly 1989a; Letton 1989; McClanahan 1989; Niles 1984; Standiford 1989; W.W. Smith 1989; Waugh 1988). U.S. census records and the testimony of Irish descendants substantiate this contention. Archival records and documents identifying the Irish masons who built the turnpike fences on particular Bluegrass farms add conclusive evidence.

By the twentieth century, the infusion of new money into the Bluegrass provided motives and means to employ crews of black stonemasons, who had learned the trade assisting Irish turnpike fencers, to construct rock fences on developing horse farms and adjacent roadways. Many fences from the turn of the century onward are attributed to this new group of craftsmen. While the work of black stonemasons remains in living memory and may thus be the source of the "slave fence" myth, these builders were not slaves.

The region's oldest fences date from the late 1700s with the establishment of plantations. The investigation of fence builders began with the origins of Kentucky settlers and the migration that brought fence-building techniques and terminology from Scotland or Scotland via Ulster. During the 1800s, many plantation owners professed admiration of the English countryside, and their desire to emulate English estates led to a proliferation of rock fences in the Bluegrass region (fig.4.1).

Fig. 4.1 PROBABLE SOURCE AREAS FOR KENTUCKY ROCK FENCE TYPES: Drystone Building Tradition (Diagrammatic)

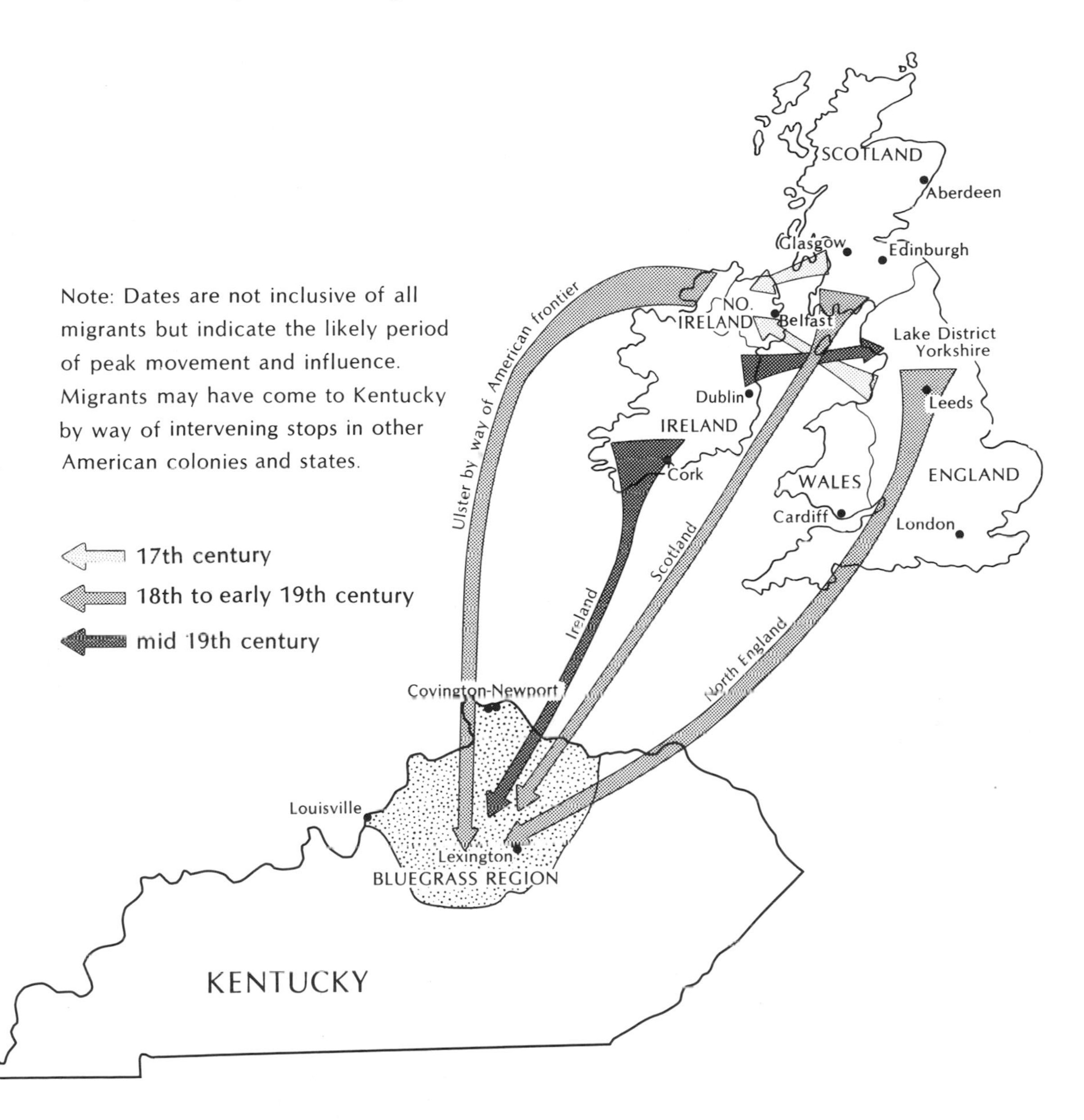

Sources: Callander [1982] 1986; Draper 12C: 26-29; Evans 1956; Holmes 1820, 395; Melish [1812] 1818; Mitchell 1880; Perrin et al. 1888; Raistrick [1946] 1988; Robinson 1984; Rollinson [1969] 1972; Shaughnessy 1986; U.S. Census.

The Irish

The Irish were widely employed building rock fences in Ireland, Scotland, England, and America during the nineteenth century. Irish geographer E. Estyn Evans relates that "a good many of [the stone walls in Ireland] are 'famine walls' built to relieve distress during famine years, in return for a meagre allowance of food and a wage of perhaps a few pence per day" (1957, 108). Groups of Irish stonemasons migrated in the mid-nineteenth century to Scotland and England, where they built enclosure walls (Shaughnessy 1986). Nineteenth-century Irish immigrants to British North America built rock fences in Newfoundland, northeast New Brunswick, and south central Ontario, Canada. Significantly, photographs and drawings show that Canadian fences match the form of Kentucky turnpike fences (Akenson 1984; Leechman 1953; Mannion 1974, 86). Irishmen worked on the rock fences and roadbeds in the Nashville Basin of Tennessee (Algood 1989) and built rock fences in southern Indiana (Leible 1952, 7; Mastick 1976, 118). Rock fences in these locations are also similar to those in Kentucky.

The potato famine from 1845 to 1848 in Ireland dramatically increased the number who emigrated to America, especially Catholics from what is now the Republic of Ireland. They migrated inland from coastal ports such as Boston and New York and often joined large construction crews that worked on canals, roads, and railroads of the eastern states (Allen and Turner 1988, 48). Although some came earlier, many of these mid-nineteenth-century Irish immigrants arrived in Kentucky, bringing masonry skills developed in Ireland.[1] Rock has been used for field boundaries in Ireland since Neolithic times and dry stone walling has been a major occupation for Irishmen for centuries (O'Neill 1984, 49). Rocky soils in the north and west of Ireland meant that farmers had to clear fields of rock and dispose of it by building fences in order to farm.

Large numbers of single young Irish men came to the Bluegrass region, where they found employment in work gangs building turnpikes and, later, railroads. Dates of emigration (listed for the first time in the 1900 census) from Ireland to Bourbon County range from 1837 to 1870, with the largest numbers arriving in the 1850s. In 1850, Bourbon County had 117 turnpikers—men whom census takers identified as road builders. Ireland was the homeland of 113 of them. Groups of Irish turnpikers, stonemasons, and laborers are likewise enumerated in 1850 in Mason, Mercer, Scott, Shelby, and Woodford counties (table 4.1). Over the succeeding decades, the number of turnpikers in each county changed as the location of major road building shifted, but the Irish dominated the Kentucky labor force until the turn of the century. Rock fence building customarily proceeded along with turnpike construction.

The observant James Lane Allen described a typical Bluegrass road-

Table 4.1. Irish Turnpiker Household, John Haley, Woodford County, 1850

Name	Age	Occupation	Birthplace
John Haley	40	Road Contractor	Ireland
Elizabeth Haley	35	Housewife	Ireland
Bridget Haley	9	------	Ireland
Ann Haley	7	------	Ireland
Patrick Haley	4	------	Ireland
Ellen O'Connell	30	------	Ireland
Patrick O'Connell	10	Turnpiker	Ireland
John O'Connell	7	------	Ireland
Bridget O'Connell	4	------	Ireland
Andrew Whelan	37	Turnpiker	Ireland
John O'Brien	30	Turnpiker	Ireland
Edward Fitzgerald	30	Turnpiker	Ireland
Daniel Drew	20	Turnpiker	Ireland
Edward Griffith	30	Turnpiker	Ireland
Datrick Dile	24	Turnpiker	Ireland
Patrick Welch	28	Turnpiker	Ireland
Richard Hannen	23	Turnpiker	Ireland

Note: The Federal Census recorder wrote "Shanty" in the margin to describe the John Haley dwelling: *Source:* Federal Census Manuscripts, 1850.

side scene of the 1890s and mentions the connection between the Irish and the rock: "All limestone for these hundreds of miles of road, having been quarried here, there, anywhere, and carted and strewn along the road-side, is broken by a hammer in the hand. By the highway the workman sits—usually an Irishman—pecking away at a long rugged pile" (1892, 30). Allen's observation suggests that Irish rock-workers had become identified with their products: retaining walls along roads, stone bridges (fig. 4.2) and trestle supports, and new rock fences on farms bordering the roadways (fig. 4.3).

One of the largest remaining rock fence collections on a single property in Kentucky is that of the Shaker community at Pleasant Hill in Mercer County. There were no slaves at Pleasant Hill. Assumption that these fences are Shaker workmanship is refuted by the fact that only one of the brethren is listed as a stonemason in the 1850 census, the first in which occupations and birthplaces are recorded, and no such craftsman is identified in 1860. Ledger and account books from 1839 to 1871, however, show that the Shakers employed many Irish stonemasons for building roads and fences. Shaker ledger books show payment of $105 in 1843 to the Irish "for turnpiking to sawmill" and in 1870 "for work on pik to the river, $30." Account books list the following employees paid for rock fence construction: Cowdry, McBride,

Fig. 4.2 TURNPIKE BRIDGE. An uncommon pointed arch bridge spans a small run of Dix River in Boyle County. Most pike bridges have round arches.

Harp, Madigan, Hoherty, O'Neal, Sears, and in some cases, simply "irishmen." A notation in 1846 of a five-dollar contribution to the Irish charity fund reveals sympathy for the plight of the Irish refugees.

The records thus clearly indicate that the Shakers themselves did not build most of their rock fences; rather, these fences are the product of Irish craftsmanship. By mid-nineteenth century, rock fences subdivided about five thousand acres at Pleasant Hill and had cost the Shakers about forty thousand dollars (Thomas and Thomas 1973, 15). In 1839 they paid Irish masons three dollars per rod of fencing; in 1841 and 1842 they paid J.M. McBride and Co. and G. and S. Harp for hundreds of rods. The Shakers were still paying Irishmen for building "stone walls" as late as 1871 (Shaker Ledger Books).

Bluegrass gentry also hired Irish immigrants to build rock fences on their landholdings. Robert W. Scott, for example, contracted with Duncan Thompson to build rock fences on the Locust Hill plantation he was developing in Franklin County during the 1840s and, incidentally, advanced Thompson $4 to join the Sons of Temperance (Clark 1990). Brutus J. Clay employed at least thirteen Irishmen between 1838 and 1876 to build rock fences on his land in Bourbon County. "Hundreds of Irish immigrants" built approximately twenty-five miles of rock fence on the Caldwell plantation in Boyle County (Caldwell 1989).

Irish stonemasons in rural areas found ample employment at their craft and, in some cases, the opportunity to acquire land for farming. John Francis Maguire, writing in 1868, advised immigrants to avoid towns: "The country is the right place for the Irish peasant, and that in the cultivation of the soil he has the best and surest means of advancement for himself and his family . . . advice I earnestly give to my countrymen" (1868, 244). One Irish Kentuckian who prospered by such advice was Thomas Barrett, who emigrated from County Cork, Ireland, to Scott County sometime between 1841 and 1850, settling near the St. Francis Roman Catholic Church in the rural community of White Sulphur. He was a stonemason and built rock fences in the Payne's Depot area of Scott County. With hard work and frugality he accumulated enough money to purchase two farms and eventually bequeathed a sizeable estate to his descendants (Bevins and O'Roarke 1985, 25; Goodman 1989).

Many Irish immigrants settled in groups in rural areas and formed ethnic neighborhoods such as the one that clustered along the Irish Ridge near Saint Nicholas Roman Catholic Church in the Eden Shale hills of western Mercer County. The McCrystal, Henry, Watters, Logue, Cox, and Huff families came together from Ireland in 1842 and settled in this locale, where each family homesteaded a small tract.[2] They were principally farmers but built rock fences as an adjunct occupation. Accustomed to rocky soils in their homeland, these Irishmen continued the practice of collecting large piles of rock and enclosing their fields as they cleared the Eden Shale

Fig. 4.3 TURNPIKE FENCES. Ledge rock fences in central Boyle County border old turnpikes.

Fig. 4.4 McCRYSTAL GRAVESTONE. "Michael/McCrystal/ May 28, 1842/Dec. 12, 1910/Mary M./his wife/ Apr. 22, 1943."

hillsides. This process continues to the present. Some of these men, John and Edward Henry, Terrance Logue, and John and Mike McCrystal, were professional stonemasons. The gravestone of Mike McCrystal, who was born "on the water" during the voyage from Ireland, is carved to resemble a coursed stone pillar, indicating his life's work (fig. 4.4) (Traynor 1989; Mrs. N. Huff 1989).

Irish oral histories in Kentucky portray close-knit extended families, with fathers, brothers, sons, cousins, nephews, and grandsons working together as fence contractors, apprenticing younger family members and passing along the skills and business for several generations. Paul Gormley, the third generation in a family of fence masons, worked as a small boy in the 1930s for his father and uncles as a "chinker" in fence construction near

Midway in Woodford County. His grandfather, Michael Augustine Gormley, came from County Tyrone, Ireland, in 1865 as a penniless stowaway. The elder Gormley emigrated in response to letters from relatives inviting him to join them in Kentucky, where they had the use of a house and plot of land. The letters told of a plain life in which planting a good garden and working hard would provide plenty to eat. Gormley took advantage of this opportunity. He prospered, married, and used his skills as a stonemason to support his family. In 1896, his sons founded the Gormley Brothers Stone Fence and Road Building Company and later expanded into the bridge building business using "only a plumb-bob and circumference circles to teach themselves engineering." They built and repaired rock fences in Fayette, Franklin, Owen, Scott, Spencer, Shelby, and Woodford counties. James Rogers Gormley, Michael's son, built his last fence in 1953 (Gormley 1987, 1988, 1990) (fig. 4.5).

The Woods family history also describes an extended family engaged in road and fence building. It depicts the transfer of fence masonry techniques to succeeding generations as well as to a different cultural group. Edmond P. Woods, who was born in 1820 in County Limerick, Ireland, apprenticed as a stonemason in his youth and came to the United States during the potato famine. Woods settled in Paris, Kentucky, where his daughter married John H. Cain, another Irishman who emigrated about 1877. Soon after Cain's

Fig. 4.5 GORMLEY FAMILY. Michael A. Gormley had six sons, all of whom were stonemasons; *back row, left to right:* John, Roger, Hugh; *front row, left to right:* Michael Jr., Patrick, Michael A. Sr., and William. *Photograph courtesy of Paul Gormley.*

Table 4.2. Irish Stonemason Household, John O'Brian, Scott County, 1860

Name	Age	Occupation	Birthplace
John O'Brian	60	Stonemason	Ireland
Ellen O'Brian	45	Housekeeper	Ireland
Sarah O'Brian	8	At home	Kentucky
Catharine O'Brian	6	At home	Kentucky
Mary O'Brian	4	At home	Kentucky
John O'Brian	11/12	At home	Kentucky
Barna Riley	35	Stonemason	Ireland
Sam Leden	50	Stonemason	Ireland
John Smith	50	Stonemason	Ireland
John Fralen	25	Stonemason	Ireland
James Shoa	15	Stonemason	Ireland
Thomas Riley	35	Stonemason	Ireland
Michael Flarrity	50	Stonemason	Ireland

Source: Federal Census Manuscripts, 1860.

arrival, Woods and Cain formed a road building and rock fence construction partnership and together contracted to build the Bourbon County section of the Maysville to Lexington Turnpike and adjacent rock fences. Four of Woods's sons and one of Cain's sons apprenticed to their fathers and assisted them in building the road and fences. The firm also employed black men as laborers. Several of the sons later went into the construction business for themselves, as did the black men on the fence crew (Miller 1989a, 1989b).

A partial list of Kentucky quarriers and fence builders compiled from census records, oral histories, published and unpublished records, and field notes, contains the names of 406 native Irish stonemasons (see Appendix 1). Census lists include the vital statistics of these immigrants. While most censuses simply note the occupation of roadbed workers as "turnpikers" or "turnpike laborers," using "stonemason" for rock or stone lay-up workers, some census takers more explicitly listed fence builders as "stone fencers" or "rock fencers." By 1850 as many as eighteen single young Irishmen all employed in fence masonry and turnpike work composed large households in Inner Bluegrass counties. The censuses show groups of stonemasons living in the same neighborhoods and together in households as the customary living pattern.[3] A common Irish household was that of a man and wife born in Ireland, perhaps with young children born in America, and also including a few single, adult men (table 4.2). Some of the adult residents had the same last name as the head of the household, suggesting that they were brothers or kinsmen. The next most frequent habitation pattern of these immigrants

Table 4.3. Stonemasons in Residence with Landowner, Junius R. Ward Household, Scott County, 1850

Name	Age	Occupation	Birthplace
Junius R. Ward	47	Farmer	Kentucky
Matilda Ward	42	Housewife	Kentucky
George Ward	18	------	Kentucky
Elizabeth Ward	16	------	Kentucky
Sarah Ward	14	------	Kentucky
Mary Ward	11	------	Kentucky
Junius Ward	7	------	Kentucky
Matilda Ward	2	------	Kentucky
H.B. Selby (female)	21	[Governess?]	Kentucky
George Laws	20	Mason	Kentucky
William McHenney	35	Mason	Ireland

Note: Mr. Ward owned 520 acres of land, and his real estate was valued at $102,000. He kept twenty-seven horses, two mules, ten cows, seven oxen, twenty-three other cattle, seventy-five sheep, and ninety swine. He raised wheat, rye, corn, and oats. He also owned seventeen slaves; ten were males, of which six were seventeen years old and older. *Source:* Federal Population, Agricultural, and Slave Census Manuscripts, 1850.

involves stonemasons living on the farms of landowners who employed them (table 4.3). Those who did not fit these arrangements dwelt with other craftsmen or tradesmen, usually of the same ethnicity.

Slaves and Freedmen

Although documents, records, and Irish oral history disprove the popular assumption that slaves were responsible for the origin and design of Bluegrass rock fences, scattered farm records and black families' oral traditions indicate that slaves were in fact part of the work force for fence building, especially on large estates. Owners of most large Bluegrass plantations also owned slaves and frequently assigned them to assist masons by digging foundation trenches and gathering and hauling the rock from quarries or creek beds to the building sites (Clay Family Papers 1836-1873; Clay and Thornton 1848; Kelly 1989a).

Fence design and construction originated with skilled master masons and experienced quarrymen who were hired and paid by the job or by the rod. Although the stonemason was, to use contemporary terms, the "lay-up man," slaves and apprentices facilitated his work. Through this process, Irish masons passed on their knowledge of tools, techniques, and terminology for rock fence building to other laborers. After emancipation, many of these laborers became stonemasons in their own right.

A letter written in 1840 to a landowner in Woodford County illustrates the relationship between slaves and stonemasons in fence building. The letter was "written by a son away from home . . . asking if the negroes were going to have enough stone hauled in for the new fence so the 'immigrants' could start as soon as the ground settled in the spring" (Kelly 1987). This letter clearly indicates that pre-famine Irish immigrants constructed rock fences, that slaves assisted in hauling material to the construction site, and that the landowner provided the building material for the masons.

Records of four Bourbon County farms—Soper's, Clay's, and the Buckners'—each of which has several miles of rock fencing, also illustrate the linkage between Irish masons, fence construction, and black laborers. Census records reveal that each of these landowners had Irish stonemasons in residence. In 1880 Irish stonemason Richard Long lived on the farm of Barton Soper in the southeast section of Bourbon County; also in residence on the farm were black farm workers, who could have assisted in building fences. Likewise, although Brutus J. Clay owned 132 slaves in 1860, he employed Irish masons to do most of the fence building on his farm with assistance from at least one of his slaves (Clay and Thornton 1848). Since freedmen often took the surnames of the families who had owned them, Horace Buckner, a black stonemason listed in the 1870 census, probably learned his skill on the farm of Walker or William T. Buckner.

Because slave censuses were separate from those of the general population and did not list the occupations of the slaves enumerated, it is impossible to identify the number of black men who may have been working as stonemasons in antebellum Kentucky. The general census lists do, however, include free blacks. The number of stonemasons identified among this segment of the population is relatively small; the 1850 and 1860 censuses combined enumerate only sixteen free black stonemasons in the entire state of Kentucky. Two examples of slaves skilled in masonry have emerged from other records. "Mrs. White's Samuel" built chimneys on the blacksmith and cooper shops of Brutus J. Clay's plantation in 1833 (Clay 1828-1846). David Thomson of Scott County moved to Missouri in 1833, taking seventy-five slaves with him, one of whom was a stonemason (Bevins 1984b).

After emancipation the number of Irish stonemasons steadily declined, while the number of black stonemasons steadily increased. Census records for 1870 list thirty-six Irish and four black stonemasons in Bourbon County and thirteen Irish and three black stonemasons in Woodford County. This proportion gradually reversed so that by the turn of the century black fence masons and turnpike workers outnumbered the Irish. By 1910 there was only one Irish, but twenty-three black, stonemasons in Bourbon County and two Irish and sixteen black stonemasons in Woodford County. Replacement of the Irish by blacks in the trade occurred in all the other Bluegrass counties (tables 4.4 and 4.5).

Table 4.4. Bourbon County Stonemasons, 1850–1910

	White					
	*Ky	**US	***Foreign	*Ireland	Black	Total
1850	2	0	0	13	0	15
1860	3	1	2	26	2	34
1870	7	1	2	36	4	50
1880	8	1	0	21	17	47
1900	5	0	0	15	19	39
1910	7	0	0	1	23	31

**Place of birth*
***U.S. birth other than Kentucky*
****Foreign birth other than Ireland*
Note: The 1850 and 1860 censuses do not list slave occupations. *Source:* Federal Census Manuscripts.

Table 4.5. Woodford County Stonemasons, 1850–1910

	White					
	*Ky	**US	***Foreign	*Ireland	Black	Total
1850	2	2	1	7	0	12
1860	10	2	0	21	0	33
1870	4	1	0	13	3	21
1880	8	1	1	19	15	44
1900	16	1	0	5	21	43
1910	3	0	1	2	16	22

**Place of birth*
***U.S. birth other than Kentucky*
****Foreign birth other than Ireland*
Note: The 1850 and 1860 censuses do not list slave occupations. *Source:* Federal Census Manuscripts.

Some black habitation patterns of the early twentieth century duplicate those of the Irish a few decades earlier. In 1900, stonemasons Eugene Watson, Samuel Guy, Berry Scruggs, and George Finex, for example, all single men in their twenties and thirties, lived in the same dwelling in the Millville Precinct of Woodford County. Elija Mason, William Smith, and Mort Dobson, black quarrymen, lived next door (U.S. Census). In some families a father such as Gren Williams and all of his sons were working as stonemasons (table 4.6).

Some freedmen initially practiced the masonry trade as employees of the Irish turnpike contractors. The road and fence construction firm of Woods and Cain of Bourbon County, for example, employed "hundreds of local black men over the years but none were slaves. In later years blacks were the only men to follow the trade as the work was heavy and non-

Table 4.6. Black Stonemason Household, Gren Williams, Scott County, 1880

Name	Age	Occupation	Birthplace
Gren Williams	57	Stonemason	Kentucky
Amanda Williams (D)	22	Housekeeper	Kentucky
Noah Williams (S)	19	Stonemason	Kentucky
Sam Williams (S)	17	Stonemason	Kentucky
John Williams (S)	14	Stonemason	Kentucky
Jane Williams (D)	9	At home	Kentucky
Anna Williams (D)	7	At home	Kentucky
Mary Wash (A)	50	Housekeeper	Kentucky

A = Aunt; D = Daughter; S = Son.
Source: Federal Census Manuscripts, 1880.

appealing to white men. After 1910 most stone fences were built by black labor" (Miller 1989b). Black fence masons became proficient in their trade and increasingly contracted work on their own.

Oral histories relate that, like the Irish fence builders, a number of black fence builders started stonemasonry businesses that remained within their families for several generations. Fence building has been the occupation for five generations of the Guy family. In 1941 John H. Guy worked on the relocation of a rock fence when U.S. Highway 60 was widened between Shelbyville and Clay Village, a fence that his grandfather, Henry Guy, had helped construct as a slave before the Civil War. John Guy's son, John, graduated from Kentucky State Industrial College and Tuskegee Institute in Alabama, then worked with his father in the masonry contracting business (Landau and Landau 1941) (fig. 4.6).[4] Guy family masons also built many of the notable mortared stone buildings in the Bluegrass during the first half of the twentieth century,[5] as well as fences on the Walmac and Elmendorf farms. The fifth generation of stonemasons worked in Bourbon County in 1959 (Thierman 1959, 44). The Guys specialize in the "random ashlar" pattern (Guy 1989). John H. Guy III is currently one of the best-known stonemasons in Franklin County, experienced both in dry-laid and mortared construction.

Stephen Lee, born 1908, and Legrand Lee, born 1893, of Midway, were members of another slave-descended family of stonemasons who were well-known for both the quality of their stonework and the repair of many of Woodford County's early fences. The Lee brothers learned the trade from their father, John Lee, who had been born a slave on the Alexander plantation. Older slaves who had learned stonemasonry from Irish immigrants working on the Alexander and neighboring farms taught John Lee (Cause 1980). The Lee family, in turn, have handed the skill on to a number of white stonemasons in practice today (Higgs 1989).

Other Rock Fence Builders

Frankfort, the state capital, was also the site of the state penitentiary by 1800. Tradition holds that prison laborers built rock fences in the surrounding countryside, but it is difficult to determine when or where. Reports of convict masons have persisted, such as a 1952 newspaper article stating that "there is one farm alone in the north-central part of [Franklin] county which has on it 17 miles of stone fence, said to be the work of no less than 200 convicts" (O'Connor 1952, 7). One possible period for this work is 1825 to 1834 when Joel Scott was keeper of the penitentiary; he is known to have directed convict labor in enterprises such as furniture making and quarrying (Kramer 1986, 83).

The James R. Holt farm in Franklin County contains fences reported to have been built with convict labor. The present farm owner surmises that the fences were built between 1859 and 1880 because Harry Innis Todd, who owned the property then, was county sheriff and seemed a person apt to have had access to prisoners to build rock fences on his farm (R. Taylor 1988). In 1905 Holt sold the farm, and a printed sales brochure describes it as

> one of the best Blue Grass farms in Kentucky, extensively and most judiciously improved, in a locality possessing every desirable requisite . . . contains seven hundred and forty-two acres of rich limestone soil, 650 acres of which is Blue Grass. . . . The land is all enclosed with solid stone fence, also all the division fences are of stone, consisting altogether of over twelve miles of fence all of which is so substantially built, that a rabbit or rat cannot pass

Fig. 4.6 GUY NAME IN FENCE. John Guy rebuilt this Fayette County fence for G.F. Willmott in 1956.

> through it; the foundation is below the frost line, and freezing and thawing have no effect on it. It will average from two to three feet across, about eighteen inches at the top, and about nine feet in height from bottom of foundation to top of fence. It is so closely and compactly built that it will last forever. The division fences cut the farm into seven fields, all of which have a spring of running water passing through them. . . . I have also one very large barn with four three acre lots in the rear, divided off by solid stone fence, very convenient for separating stock when desired. [Holt 1905]

While very well built, the remaining fences are not of the dimensions described; to be nine feet in height they would have to have been buried four feet in the ground. Holt's sales brochure does, however, demonstrate what progressive farmers considered desirable farm enclosures: "all the land"; division fences enclosing fields, each containing spring water; and four three-acre barn lots with rock fencing separating each lot from the others. Rock fences were seen as ideal enclosures and as symbols of virtuous agriculture in the early twentieth century.

The circa 1810 report of traveler John Melish further complicates efforts to establish the period when rock fences might have been built by convict labor on the Holt farm, by providing another instance of convicts doing stone work: "The [Kentucky state] penitentiary is under such excellent management that the institution supports itself by a judicious application of the labour of the convicts. . . . Various mechanical branches were carried on; but the convicts were mostly employed in sawing marble in the open yard" ([1812] 1818, 398). While sawing marble (Oregon limestone) is not the same as building rock fences, this report does show that prisoners were working with rock, at least in the yard, as early as 1810.

Oral tradition indicates that two additional cultural groups participated in building rock fences in isolated cases. About 1920, Mexican masons built the two and one-half miles of road frontage fences at Xalapa, a famous Bourbon County horse farm (Waugh 1988; Mogge 1989). French-speaking blacks from the Caribbean islands with the family names of Meaux and Trumbo built rock fences in Boyle County. They lived in a settlement on Pope Road, in small stone houses surrounded by rock fences (Caldwell 1989). The 1860 census of neighboring Mercer County, which lists black stonemason Robin Meaux, then fifty-eight years old, supports this information.[6]

Sources of Kentucky's Oldest Rock Fence Traditions: Ulster, Scotland, and Northern England

While the immigration of experienced Irish stonemasons to Kentucky during the famine years of the mid-nineteenth century and the subsequent development of the fence masonry trade within the black community after the Civil War accounts for the rock fences built in central Kentucky after the mid-1800s, these two groups of masons were not responsible for rock fences built earlier. Although rock was not the dominant fencing material in the Bluegrass until after 1840, rock fences were built as early as 1777. Oral history maintains that some of central Kentucky's rock fences are as much as two hundred years old, and scattered public records and written accounts verify this (Draper 12C:26-29; Perrin 1882b, 575; Kramer 1986, 53; Everman 1977, 25; Peet 1883, 54). What was the origin of these fences, and how had they come to be built in Kentucky?

No documents have surfaced that specifically identify the builders of Kentucky's late-eighteenth- and early-nineteenth-century plantation fences. Piecing together circumstantial evidence provides only an informed guess that points to Scotland or Scotland via Northern Ireland (Ulster) as the source.

Rock fences—called stone ditches, stone dykes, stone fences, or stone walls, depending on local terminology—are traditional in western and northern Ireland, Scotland, northern England, and Wales. Dry stone walls were built by the Celts, ancient inhabitants of all the North Sea's great islands (Vince 1983, 85; Rollinson [1969] 1972, 18).[7] The practice of dry stone walling in the British Isles is thought to have originated in Scotland in the "stone age" (Mitchell 1880, 75). In the north of Ireland "the art of dry walling is of megalithic antiquity, and many examples of walls of prehistoric age lie buried under the peat" (Evans 1956, 16).

Most writers agree that the oldest extant stone fences in the British Isles date to the early 1600s, by which time stone was replacing other materials for enclosures in Scotland and northern England. By the 1700s, dry stone fences were the norm in Scotland, northern England, and Ireland (Callander [1982] 1986; Raistrick [1946] 1988; Rollinson [1969] 1972; Evans 1956), and the vast majority of existing walls were constructed during the eighteenth and nineteenth centuries (Brooks 1977; Evans 1956; Mannion 1974; Raistrick [1946] 1988; Rollinson [1969] 1972).

The north of Ireland was the ancestral homeland of a large proportion of the first settlers in the Kentucky Bluegrass region (Wilson 1933)[8] and most of these families were descendants of people from Scotland and northern England (Murray-Wooley 1988), where stone walls were customary. Did the tradition of using rock for fence building transfer from Scotland to the north of Ireland? Did it transfer from there to regions of America where Ulstermen settled before they came to Kentucky?

The Glass, Colvin, and Vance families, all of Ulster stock, built rock fences in the late 1700s in Frederick County, Virginia, the heart of the Shenandoah Valley (Hofstra 1988). These fences exactly match the form of those built in Kentucky. In fact, similar fences are scattered throughout the

Valley—in Pennsylvania, Maryland, and Virginia—in areas settled by Ulstermen. The pervasiveness of these artifacts in areas of strong Ulster settlement suggests that this fence type was likely built by Ulstermen in the middle colonies. Further, the Valley fences differ from New England's stone walls, which were built by people of English descent. These differences, however, have as much to do with the characteristics of local stone as with regional construction techniques (fig. 4.7) (Fields 1971).

During the last quarter of the eighteenth century, settlers from the Valley of Virginia moved into Kentucky. Many of the more prosperous among them built substantial stone dwellings on their new plantations in the Bluegrass region. A comprehensive study of stone house sites in Kentucky reveals that the majority of these farms that contain rock fences originally had Ulster-descended owners. The incidence of rock fences at stone house sites is significantly higher among Ulstermen and Scots than among any other ethnic groups.[9] This finding suggests a cultural affinity for this fencing choice in the Bluegrass even before scarcity of timber became a factor.

Fig. 4.7 NEW ENGLAND FENCE. A break allows rounded pebbles to spill from the fence interior. It is very difficult to construct a coursed fence with tie-rocks using glacier-worn material. In some areas of New England, mica schist and granite are split for fence building. *Photographed by Neal O. Hammon.*

Prior to the influx of southern Irish masons throughout the British Isles in the nineteenth century, tenant farmers themselves often erected rock fencing. Less prosperous farmers in the Bluegrass also built their own rock fences, demonstrating that this skill remained within the settlers' repertoires. Rock fences marking the boundaries of the Littral farm on Boone Creek in Clark County, for example, were personally built by the Ulster ancestor of the present owner (Creger 1989). On higher-priced land it was more common for planters to hire masons for fence building. Whether rock fences were built by the farmer or an employee, during Kentucky's settlement period landowners and landless laborers were largely of the same ethnic stocks and would have carried the same traditions.

While the prevalence of rock fences at sites first owned by Ulster descendants seems to suggest Ulster as a source for Kentucky's earliest rock fences, tying these fences to Ulster traditions by the circumstances of land ownership is insufficient for establishing the origin of the fences in the Bluegrass. In fact, searching for prototypes for Kentucky rock fences along these lines is problematical for several reasons. The precise origin of this fencing technology is obscured by migrations and acculturation of ethnic groups within the British Isles prior to migration to America (DeBreffny 1982; Lehman 1980; Robinson 1984). Furthermore, expert cultural historians in Northern Ireland cannot identify construction differences between fences in Irish and Scots settlement areas, other than those that can be explained by rock structure: "Our stone dykes are generally restricted to poorer, marginal lands where field stones are disposed of by building fences with them. For most of the *cultivated* landscape this does not apply in Ulster. Also the geology and morphology of these field stones do not lend themselves in Ulster to sophisticated construction. We do however have some later ones, with two almost independent walls with a rubble core" (Robinson 1990). These later fences, although similar to turnpike-type fences constructed in Kentucky during the second half of the nineteenth century, cannot be considered as prototypes for plantation fences because they postdate Kentucky settlement.

The old dykes of Scotland, however, are morphologically identical to the earliest Kentucky rock fences and, dating for the most part to the Enclosure Acts of 1710 (Rainsford-Hannay 1957, 23), are old enough to be candidates for prototypes. Because Irish of Scottish extraction (the largest cultural group in Northern Ireland) and Scots composed a significant proportion of the late eighteenth-century Kentucky population, speculation that the fences have a Scottish origin is warranted.

There are recorded examples of Scottish fence masonry in America. Hugh Holmes of Winchester, Virginia, described his fences in detail in an 1820 letter and explained that he was having his nine-hundred-acre farm enclosed with a stone fence by a Scotsman who "came over about two years since, and is now living on my land" (Holmes 1820, 395). Oral tradition holds that before the Revolutionary War, Scots constructed stone fences like those in Kentucky in Dutchess County, New York (Colbert 1980). Before the mid-nineteenth century, Scots built similar dry stone walls at Galt in Ontario and carried their craft as far as Australia, where Scots and Yorkshiremen built dry stone walls in Victoria (Rainsford-Hannay 1957, 92).

Several known Scottish stonemasons and quarriers worked in Kentucky. Masons whose lives spanned the mid-nineteenth century include John William Carmickle, James Fairweather, Anderson McClane, John Renan, and George Harrison Waugh. Descendants of two Kentucky families, the Waughs of Scott County and the Carmickles of Mercer County, have continued Scottish fence-building traditions to the present. These fences differ structurally from those built by the Irish.

Characteristics of Irish fences built in Kentucky that separate them from Scottish fences are similar to features of Irish fences noted in the British Isles. Scottish writer Frederick Rainsford-Hannay commented that Irish walls have much less batter than Scottish walls and do not contain throughbands (1957, 83, 84). He noticed in some Irish walls the tendency to place the bigger stones "lengthways, along the [face of the] work and not, across the wall" (84). He also reported that Irish walls can often be identified by a coping of mortar, while others are finished by having the small stones loosely placed on, not built into, the top (84). Typical turnpike fences in Kentucky have all of these traits, as do many twentieth-century dry-laid fences.

The most convincing argument for a Scottish source for Kentucky's earliest rock fences are published explanations and how-to drawings of the stone walls in Scotland and northern England (Dry Stone Walling Association 1989), which depict structures identical to plantation fences in the Bluegrass. These images are so similar to Kentucky fence morphology that quoting them seems repetitious, yet the parallels in terminology and technique provide evidence that Kentucky's plantation fences derive from Scottish building traditions.

Perhaps the best-known British publication describing these walls was written in 1957 by Rainsford-Hannay.[10] His book, *Dry Stone Walling*, includes sectional drawings and instructions for correctly building a wall:

> The foundation trench would be dug out to the firm subsoil; six inches is generally deep enough. . . . The width of the trench is at least 4 inches wider than the base of the dyke. . . . Foundation to be 32 inches wide and the base 26 inches wide . . . from there build to taper [batter] gradually to 14 inches wide at the top . . . both sides brought up together, having the stones properly hearted and

> packed in the centre . . . the stones to lie with their ends [length] inward, so as to stretch into the dyke as far as possible, for the better binding of the work . . . the principle of "breaking joint" being followed as far as possible. . . . Height to be 4 feet 6 inches . . . throughbands [tie-rocks] 21 inches above the grass at 1 yard centres, projecting slightly on each side . . . cover band stones [cap course] on top to project 2 inches on either side. [1957, 35-37, 41]

The only difference between these walls and the fences built in Kentucky is the preferred direction of the slant of the coping—uphill instead of down—although Rainsford-Hannay mentioned walls in Cumberland with downhill-slanted copings similar to Kentucky examples (1957, 77).

The twentieth-century dyking work of Charles Scott Jardine, of Dumfries, Scotland, whose family contains generations of stonemasons, fits Rainsford-Hannay's report. Jardine additionally recommends two rows of "throughs" (tie-rocks) positioned at fifteen and twenty-seven inches from the ground (O Borchgrevink 1982, 83). This pattern appears in Kentucky on fences such as those at the Buckner farms in Bourbon County. Jardine also builds a cap course below the coping that projects two or three inches from the face of the wall (O Borchgrevink 1982, 84), another detail found in Kentucky fences.

Explanations of walling techniques in Yorkshire and the Lake District of England are consistently similar to those across the Scottish border. Wallers in the West Riding stress the importance of binding the two sides of the wall together with through stones, or "binders," and of firmly packing the filling, or hearting, to prevent slippage or subsidence (Brooks 1977, 101). North British wallers pride themselves on judging by eye where and how a stone will best fit (Rollinson [1969] 1972, 7). A Yorkshire craftsman asserted that "walling requires a keen sense of spatial relationships and it is a matter of pride to wallers to estimate so well that it is not necessary to pick up any stone more than once." They also feel that it is a matter of skill not to use a hammer, or at least to use one as seldom as possible (Brooks 1977, 98-99). John G. Jenkins advocated leaving the stones whole: "The true stone waller does not cut his raw material to size if he can possibly help it" (1965, 169).

Dry-laid flat-coursed plantation fences in Kentucky exhibit all of these characteristics of Scottish and northern English dry stone dykes and walls, features that distinguish them from fences of Irish origin. Scottish-descended fence masons in Kentucky continue to use Scottish construction techniques. These include leaving most rocks whole, laying the longest length of a rock into the fence, carefully packing the inside as they build, and placing tie-rocks entirely through the fence every few feet (Carmickle 1990; Waugh 1988). Kentucky masons also take pride in their ability to judge on one try where a rock will best fit.

The possibility remains that Scottish walling practices were transferred to Irish laborers on Kentucky construction jobs. As early as 1803 a crew of thirty "Irish" stonemasons worked for contractor Peter Paul in Fayette County. Paul's father was an Ulster-Scot. Paul may have taught and required his crew to use the masonry techniques to which he was accustomed and of which he approved. In this way, Irish laborers could have assimilated Scottish walling design and carried on its use. But Paul's "Irish work crew" may not be the clear ethnic identification it seems. During the early nineteenth century in Kentucky, immigrants from the north of Ireland, the Ulster-Irish, were referred to as "Irish." It was not until the incursion of substantial numbers of Irish Catholics in the second quarter of the century that Ulstermen came to be referred to by the misnomer "Scotch-Irish."

Further, the Irish may have learned Scottish construction techniques in addition to their own before they immigrated to Kentucky. Not only were Irishmen employed on walling projects throughout the British Isles, but Scottish masons residing in Ulster may have worked with Irish masons there. It is therefore unclear whether the majority of Scottish-type fences in Kentucky—the plantation fences—came directly from Scotland or were brought from the north of Ireland by Ulstermen of Scottish and northern English descent.

Construction details of Scottish and Kentucky plantation fences, the time in which they were built, and the migration of people of Scottish ancestry to Kentucky indicate Scotland as the source of these oldest rock fences in the Bluegrass. While the evidence supporting this attribution is in part circumstantial, it is nevertheless persuasive.

Additional Comparisons: Kentucky, Scotland, England, and Ireland

Changes in the structure of Kentucky's rock fences that emerged during the course of the nineteenth century and became commonplace in the twentieth also occurred in stone walls of the British Isles, providing additional parallels in the history of the craft in both places. Northern British wallers criticize some of these techniques.

One disapproved practice in Scotland is hammering small pieces of stone, called "pins" in Europe and "chinking" in Kentucky, into the wall joints to fill the spaces. Rainsford-Hannay warned that "when pins are tapped in to fill in interstices, they must inevitably slightly alter the tight bedding of the bigger stones" (1957, 41). Another writer stated that pinning is a symptom of laziness in building and that it forces the stones out of alignment (Brooks 1977, 104). In some regions this procedure was thought to have tightened the wall, and builders finished walls by wedging in small stones with a stone hammer wherever room could be found for them (Jenkins

1965, 171). Such chinking is customary in Kentucky turnpike fences, but not in the earlier plantation examples.

A second trait of modern British Isles dykes and walls that occurs in Kentucky turnpike fences is hearting (filling) that is not tightly packed. In Scotland, Charles Scott Jardine reported that he always admonished apprentices not simply to throw in the hearting as might be easiest (O Borchgrevink 1982, 84). Fill that is shoveled in washes out easily, offers no internal structural support, and is "worthless" (Brooks 1977, 101). Rainsford-Hannay observed that "a well-built dyke always looks well, but a badly hearted dyke can look well too. At all costs the temptation to show a good outside at the expense of a poor hearting must be avoided. . . . Only good hard stones should be used, nothing small enough to get washed out by rain. It is fatally easy to take a shovel full of sand or soil here and there to level up the hearting and to make what looks like a good bed for bigger stones. But this means death to the dyke" (1957, 40). Kentucky fences built for landscaping value rather than for confining stock often have such loose fill.

In a book published in London in 1877 Henry Stephens criticized stone walls in England constructed more for aesthetics than function. He felt that they were built on "erroneous principles, the stones being laid more with a view to make a smooth face than give a substantial hearting to the wall." He attributed this neglect to "ordinary masons [as opposed to wallers] who, being accustomed to the use of lime-mortar, are not acquainted with the proper method of bedding loose stones in a dyke as firmly as they should be, and are therefore unfitted to build such a dyke" (478). After cement mortar became generally available in the late nineteenth century in Kentucky, masons experienced in the use of mortar for stone veneer work sometimes engaged in dry fence construction, often with structural results that were less than ideal.

An additional similarity between English and Irish walls and Kentucky turnpike fences relates to the practice of placing the longest dimension of the rocks along the outer face of the wall instead of perpendicular to it. This practice is known in Britain as "trace-walling" (Simkins 1989c). A British writer today notices with irony that thousands of tourists admire the mellow stone walls of the Cotswolds each year although walling experts from other parts of Britain regard them as structurally inferior. Large stones are difficult to obtain in the Cotswolds; thus, heavy copestones and functional ties are rare in this area's fences. In spite of the scarcity of large stones, "it is difficult to defend the Cotswold waller's inclination to lay stone with their length along the wall rather than into it" (Garner 1984, 22). Rainsford-Hannay observed this same feature in Irish walls (1957, 84). Not only are these construction details temporally diagnostic in Kentucky, they also help identify fences built by masons of particular cultural heritage on both sides of the Atlantic.

Possible English Influence

Since England was also the ancestral homeland of many of Kentucky's first settlers it is appropriate to question whether English stone walls were possible models for Kentucky's plantation fences. No reliable research about the numbers or specific origins of Kentuckians of English heritage is available, making any stone wall building tradition from England difficult to trace. This problem is not relevant to identifying prototypes for Bluegrass fences, however, because most English walls postdate the first Kentucky plantation fences. Stone walls did not exist before 1760 north of the Thames and were "fairly complete only by 1850" (Rainsford-Hannay 1957, 20). Most of the stone walls of the Lake District and Yorkshire are "relatively recent," built in response to the General Act of Enclosure of 1801 (Rollinson [1969] 1972, 24; Raistrick [1946] 1988, 9), by which time the Bluegrass region of Kentucky was already settled, with some plantation rock fences in place. The fact that nineteenth-century walls of the English Cotswold region are newer than and are structurally unlike the plantation fences of Kentucky indicates that they are not possible prototypes of the Kentucky fences.

Admiration of the English landscape, however, appears to have been one factor influencing the construction of Kentucky's mid-nineteenth-century turnpike fences. In the Bluegrass, "a purely English taste was shown for woodland parks with deer and, what was more peculiarly Kentuckian, elk and buffalo. . . . There was the English love of lawns, with low matted green turf and wide-spreading shade-trees above . . . the English fondness for a mansion half hidden with evergreens and creepers and shrubbery, to be approached by a leafy avenue, a secluded gateway . . . a winding stream, an artificial pond, a sunny vineyard, a blooming orchard, a stone wall" (Allen 1892, 61).

This was an acquired taste. Several of the most socially and economically prominent gentlemen farmers of central Kentucky descended from families that had lived for several generations in areas of Virginia where there were no stone fences. By the nineteenth century many of these men had traveled to England, however, to attend the universities and to purchase blooded stock. They admired the beauty of the English countryside and were inspired to create replicas of English country estates, which included newly built enclosure walls, on their Kentucky plantations.

Chaumiere du Prairie in Jessamine County was perhaps the most famous and influential landscaped estate of the nineteenth-century Bluegrass. It was established by David Meade, grandson of an Irish Catholic from County Kerry. Meade, who inherited most of his father's estate in Virginia, was a man of large fortune and had been educated in England (McDowell 1981, 111). He moved to the Bluegrass in 1796, where he purchased three hundred acres and developed an extravagantly landscaped park: "He imported rare and exquisite plants. He made lakes, constructed water falls, shaped islands, built summer houses and porters' lodges . . . and maintained in every way the style of a feudal lord" (Young 1898, 217). About 1814, when Dr. Horace Holly of Transylvania University visited Chaumiere, he wrote that Colonel Meade "has been a good deal in England in his youth, and brought home with him English notions of a country seat. . . . Everything is laid out for walking and pleasure. The whole is surrounded by a low, rustic fence of stone" (Peet 1883, 55). If the English countryside had inspired Colonel Meade, Chaumiere surely inspired his visitors and wealthy Anglophile neighbors, to which the construction of mid-nineteenth-century rock fences on neighboring estates is testimony.

Many influential gentlemen farmers from the Bluegrass traveled to Europe. Brutus J. Clay, Robert Aitcheson Alexander, Robert W. Scott, Charles T. Garrard, and Lewis Sanders were among the members of the Kentucky Agricultural Society whose sojourns in England may have affected their ideas of admirable and imitable landscape design, including rock fences.

CHAPTER 5

Brutus Clay's Auvergne

What . . . can please the eye of a countryman more than the unbroken verdancy and fertility of a Kentucky blue-grass farm?

—John Burroughs, *Riverby*

Few nineteenth-century family farm records survive that provide detailed information on farm establishment and management. One of the most complete collections of personal correspondence and business records for a central Kentucky farm is the Clay Family Papers.[1] The papers document the day-to-day affairs of Green Clay and his son, Brutus J. Clay, of Bourbon County and illustrate the rock fence building process on a large Inner Bluegrass farm. The Irish stonemasons and quarrymen who built many of the region's antebellum fences are central figures in the records, which also illustrate the process whereby slaves and freedmen "apprenticed" in stone-masonry with Irish masons. Moreover, because Brutus Clay kept meticulous accounts and recorded his business affairs from the first years that he began to develop his farm, these papers provide an overview of the long-term settlement process. Contracts and account book tabulations record the problems associated with wooden rail fencing and how the shortage of timber might have affected the decision to build fences of rock.

The Site

Bourbon County lies along the eastern edge of the Inner Bluegrass and has been known as a fruitful land since the early years of settlement in the 1780s. The village of Paris, the Bourbon County seat, grew up where the road from Maysville to Lexington crossed Stoner Creek. Paris is central to an extensive rolling plain that attracted settlers. The land was covered with cane, clover, and great hardwood trees, all signals to the discerning eyes of surveyors and settlers that fertile soils lay below. Many Virginians received French and Indian War or Virginia Militia service land grants in the Kentucky Bluegrass. Some sold their patents but others moved their families and livestock to central Kentucky.

Among the migrants was General Green Clay who moved from Virginia to Madison County near Richmond, Kentucky, coming to claim a large tract of land. Clay was the fifth American-born descendant of John Clay of Wales, who died in Charles City, Virginia, in 1638. He was named Green in honor of an ancestor and was related to the Mitchells, Wilsons, and Frames. Clay served in the Revolutionary War and, in the 1780s, in the Virginia House of Delegates. In 1805, his second son, Brutus Junius Clay, was born in Madison County. Educated at Centre College in Danville, Brutus Clay moved to Bourbon County in the 1830s and, although only a young man in his twenties, began to accumulate land. By the eve of the Civil War he had assembled one of the largest estates in central Kentucky (Perrin 1882b). Brutus Clay's land included some 800 acres between Kennedy Creek and Stoner Creek that his father, Green, had received as a Virginia patent in 1783.

Young Clay was a gentleman farmer, not in the sense that he was a high-born squire of title, but because he was educated and studied farming as an avocation at the same time that he depended upon the profits from his stock and crops to provide sufficient income to sustain his family and lifestyle. Clay subscribed to agricultural periodicals and bought books describing the latest ideas in crop and animal husbandry. He actively supported improvements in stock breeding. Together with several friends, he organized the Northern Kentucky Cattle Importing Company in 1853, which pooled investors' money and sent representatives to England to purchase the best breeding stock they could afford (Garrard [1853] 1931, 400).

Perhaps because of interest, or perhaps because he acted as his family's financial officer and was responsible for administering several estates, Clay was an enthusiastic record keeper. He listed day-to-day events associated with farm and household management in ledger and memorandum books. He issued receipts and made copies. He drew up contracts for a wide variety of jobs and services that he hired others to provide. The records document the building of the farm's rock fences, many that still stand.

Clearing and Fencing the Land

Land without fences was of marginal use for agriculture in nineteenth-century Kentucky, regardless of how fertile it might have been. This concern appears early in the Clay farm records. Green Clay owned several different land patents in Campbell and Estill counties, in addition to his homeplace, Claremont, in Madison County and the property south of Paris in Bourbon County that he deeded to his son Brutus. He leased or rented small acreages to tenants without a fee but required that they improve the land in return for the crops and stock they could produce farming it. Clearing and fencing the land was the most common improvement Clay required his tenants to make. A lease agreement between Green Clay's attorney, Stephen D. Lewis, and tenant John Yeager describes the process:

> I John Yeager hereby lease of Stephen D. Lewis the plantation where I now live for four years to commence the first day of January 1820. and oblige myself my heirs to clear and inclose ten acres of land fit for cultivation by cuting all timber one foot over and under down and deading all other trees and deliver said place to said Lewis or his agent or attorney, at the end of the year 1823. said ten acres is to include what is now cleared on said place and the fence to be five feet high when delivered all round said ten acres and said John Yager to have the place Rent free on the terms afore said Yager not to cut or Destroy any timber in waist no sell none on the land all houses and other improvements to be delivered at the end of this lease without waist or distruction witness our hands and seals. This 28th day of July 1819 for the true and faithful performance of our perspective parts we bind ourselves our heirs etc.
>
> Stephen D. Lewis, Trustee by his agent and attorney in fact
> Green Clay. John Yager [Clay and Yeager 1819]

This contract illustrates the careful consideration given to clearing fields and pastures, fence building, and preserving the remaining timber. It also demonstrates a linkage between traditions of Old World Britain and nineteenth-century Kentucky, for it was common for British and Irish landlords to stipulate the type and quality of fencing that tenants were expected to build on the lands they occupied (Evans 1956, 14).

The Clays were also concerned that all green timber not be cleared. This is apparent in a 1825 agreement: "Wyatt is not to cut Green Timber out of his clearing unless said timber sufficient to make the fences cannot be got in the clearing in that case Green Timber may be cut to make up said fences" (Clay and Wyatt 1825).

The contracts illustrate two concerns that Green Clay had in the clearing and enclosing process. First, Yeager's fence had to be at least five feet high so that it would be legal as defined by Kentucky fence law. Second, Clay carefully instructed his tenants as to which timber could be cut for rails and which was to be left. Two interpretations of this point are possible. By 1825, central Kentucky farmers were beginning to experience a shortage of timber appropriate for fencing material, and Clay's caveat may have been intended to preserve those valuable stands that remained. Alternatively, the explicit instructions to preserve standing timber could have meant that Clay wished to retain the park-like character of his land, as was customary, anticipating the usefulness of mature trees in a pasture as summertime shade for grazing cattle (fig. 5.1).

Not all rail fences were built to the same standards, and the quality of farm fencing became a measure of the occupant's care and talent, no matter

Fig. 5.1 WOODLAND PASTURE. Old hardwood trees stand in one of Auvergne's woodland pastures, enclosed on three sides by rock fences. The block-like fencing material came from the farm's quarry about three-fourths of a mile away.

how limited his other abilities might be. A letter written to Green Clay by Martin Davis, one of Clay's tenants, in January 1825, expresses Davis's concern that his landlord appreciate his efforts and view him as a diligent farmer: "We are all well and at work on our fence as hard as we can I have got about 3 hundred paniels done ten rales high I am determen to have the best fence in this nabourhood before I quiet" (Davis 1825a). Three weeks later, Davis wrote to Clay of the condition of his wintering stock and the progress made clearing and enclosing land. This second letter from Davis makes clear that small fields are being consolidated into larger ones:

> There is five lambs the dogs and wolves have not found them out yet the stears are in tolerably good order are gaining flesh smartly and I have 12 head of cattle and I come on tolerbly well with my work I have put up fourteen hundred and eighty panels of fence and have stratened the fence in ever direction and taken in all the old fields and cleared land that was lieing idle. . . . I gave the letter to M Ward [another tenant on a small acreage whom Green Clay had asked to leave] and he requested me to let him stay until he could regain himself I have just commenced halling the rails from the place where Ward lived to the Quick field to repair that I do not think it possible for to finish all the fences on both sides of the Creek . . . I do not expect to finish the upper side of the Creek unless the weather is very favorable more than the out side fence to secure the crop. [Davis 1825b]

Clay's sons adopted their father's strategy for preparing his extensive landholdings for occupancy and full-scale commercial production. In 1829, young Brutus Clay leased a twenty-acre tract to Richard Wyatt for three years. Wyatt agreed to clear and enclose the land with a good fence ten rails high and to use long-lasting timber in its construction (Clay and Wyatt 1829). In this manner, tenants cleared and fenced much of the Clay property in central Kentucky so that when the owners moved onto the land the basic improvements required for operation as a livestock farm were to some extent complete. This use of tenants to clear, fence, and farm land during the period from 1790 to 1830 was an important mechanism in the transition from land held in large and relatively undeveloped blocks used primarily for open pasture grazing to farm-sized units that produced a variety of agricultural commodities.

At some time in the 1830s, Brutus Clay, an active participant in this transition process, moved to his Bourbon County property and began to build structures, lay out roads, organize fields subdivided by fences, plant a garden and orchard, and educate himself in the fine points of creating a functional and productive farm. By 1831, a list of Clay's taxable property included 210 acres on Stoner Creek in Bourbon County, 800 acres on Licking River, and over 1000 acres in Estill County. The Stoner Creek farm became the core of his home plantation, later named "Auvergne."[2] Since Clay was a large slaveholder, slaves performed much of the daily farm work such as field preparation, planting and harvesting crops, as well as tending livestock. Viewing these activities as routine, Clay may not have felt it necessary to record them in detail in his Day and Memorandum Books.

While he developed his estate, Clay lived in a rather simple log house near the center of what was to become the farmstead. Memorandum Book entries in 1830 and 1833 indicate that he had hired a slave named Sam from Mrs. White to build chimneys for the blacksmith and cooper shops under construction (1828-1846). A fashionable brick house for which Clay drew the plan was under construction by 1836 (fig. 5.2). Clay signed an agreement with Abraham Rice, brick contactor, to build this house, and hired Greene Dejarnatt and his carpentry crew to do the joinery and finish work (Clay Family Papers 1837). That same year Clay hired Andrew Vinson and Peter Terrin to quarry rock and make stone sills for the house (Clay 1828-1846).

Limestone bedrock lay below several feet of earth across much of Clay's farm so that rock could only have been quarried in hillsides where it cropped out near the surface. No evidence exists today that rock was removed from anywhere other than the large walled-in quarry east and north of the farmstead (fig. 5.3). There, about two-thirds of a mile from the house, Clay's quarrymen and masons opened a large quarry in the Lexington limestone. The rock lay in horizontal beds ranging from a few inches to two feet or more thick.

Fig. 5.2 AUVERGNE MANSION. Two lions carved by T.L. Harkins, master stonemason, guard the entrance to Brutus Clay's house. Rock fences enclosed the yard, cemetery, adjacent orchard, livestock lots, and pastures.

Fig. 5.3 EXISTING ROCK FENCES ON BRUTUS CLAY'S AUVERGNE FARM

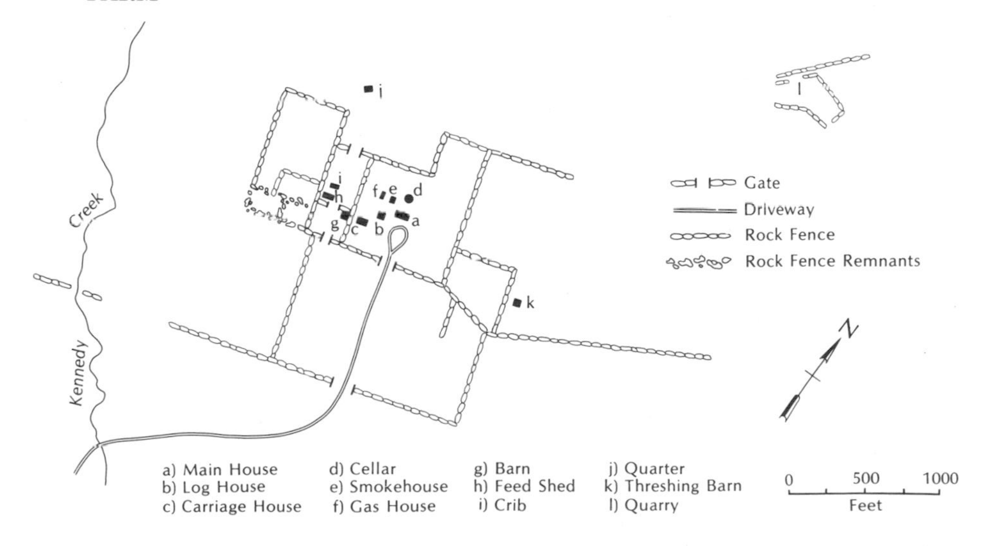

Fig. 5.4 AUVERGNE QUARRY. Having the Augergne quarry in a low draw between hills allowed quarrymen to easily remove the horizontal rock layers from the hillsides in each direction. The quarry floor was nearly level when it was actively worked. Masons walled the quarry sides as retaining walls, probably to stabilize the working face and prevent loose rock from giving way under the weight of large cattle. This is the largest known rock fence quarry in the Bluegrass, covering about one acre and extending well beyond the left and right edges of the photograph.

The quarry lies in a depression between low hills so that, as workmen removed material, they simply followed the beds into the surrounding hillsides, exposing more rock (fig. 5.4). The quarry was worked for over thirty years and eventually grew to occupy almost an acre. Though shallow, its highest wall spans over eight feet from the floor to the level of the surrounding field. On the north end, two ramps lead from the quarry floor up to field level. Although the grade is steep, the ramps allowed wagons or sleds loaded with heavy rock to move out of the quarry with relative ease. Like the working faces, the ramps are walled with rock in the fashion of the walls that were eventually built to enclose the stock lots around the farmstead (fig. 5.5).

The construction of Brutus Clay's home proceeded. In 1837, Rice and his helpers worked from February through September. From May through October, Andrew Vinson, James Boon, and James Yates did stone work on the property—sills, columns, steps, flagging, and hearths. In May, Clay ordered nails and 60,000 shingles and contracted a painter. By August, the copper gutters and spouts were installed on the roof (Clay Family Papers 1837).

Building Rock Fences

While work on farm buildings proceeded, Brutus Clay continued to develop his acreage. A receipt written in March 1841, to Sam Chew, Esquire, lists payment to the tenant Chew for improvements made to a farm rented from Clay. Clay also credited Chew for "halling and putting up 12954 old rails at .25 per hundred" (equals $32.371/2) and for "making and putting up 1329 new rails at $1.00" (equals $13.29) (Clay 1841). Three stonemasons—Abraham Rice, Francis Thornton, and Jesse Barlow—worked for Clay in 1839. Entries in the Memorandum Book for 1838 and 1839 list payments for stone work but do not specify the task. In September of 1843, Clay paid Patrick McGory $20 for one month of quarry work. This note on McGory is the second indication of Clay's employing Irishmen for stonemasonry work. Francis Thornton, an Irishman from County Cork,[3] received $35 for rock fence work during the year, and James Boone built stone foundations for Clay's lumber house and dairy, for which he received $42 (Clay 1828-1846).

The records for 1844 contain details about masons working on fences. Clay drew up an elaborate agreement with Francis Thornton to construct a rock fence on a road at the back side of his farm:[4]

Fig. 5.5 QUARRY RAMP. Two long ramps provided a relatively easy route for wagons and sleds into and out of the quarry. The ramp's sides are faced with coursed rock without a coping.

> Articles of agreement between Brutus J. Clay and Frances Thornton (to wit) the said Thornton to quarry the rock and put and complete the stone fence on Clays farm on the Paris & N. Middletown road, from where he left off last fall to Clays corner in said road next to James Neil & Spencer, of the same dimensions as the portion now completed, to wit. 28 inches at bottom 18 inches at top five feet high above ground including the copping & at least six inches in the ground. Clay to hawl the rock & he to assist in loading & c. for which Clay is to pay him forty Dollars when completed & c____ which is to be completed this fall. It is also agreed between the parties that Clay shall pay them one Dollar & twenty cents per rod for all the fence they shall put up in like manner at the other end of said string of fence beyond Parish house, Clay to furnish the rock on the ground or two Dollar & 25 cents per rod for all the fence they quarry the rock for and build. no work to be paid untill the fence is completed. the work is to be as well done as the part now finished, in good workman like manner, with binders running through the wall at proper distances so as to make a good strong fence four feet and a half above ground April—1844—Clay to dug out the foundation. this Francs Torton Mark [Clay and Thornton 1844]

This contract provides insights into the specific aspects of rock fence construction including dimensions and the use of binders or tie-stones and coping. Other records from 1839 show that Francis Thornton had been employed by Clay as a stonemason, and from this 1844 agreement it is clear that fence construction at Auvergne began before that year. Clay had hired Irishman William McGory to quarry rock at his Stoner Farm in 1843, but this record does not specify who provided the rock that Thornton used for fence building. Clay's slaves assisted in quarrying, hauling rock to the building site, and piling the rock next to the fence line; they may have also dug the six-inch-deep foundation trench for the fence.

The document explicitly mentions fence dimensions: the base was twenty-four inches. The four-and-one-half- or five-foot height requested was typical for fences of the day and constituted a legal fence. Clay's concern for a quality fence suggests that he had become knowledgeable in the nuances of fence construction. Placing the base course below the frost line added to the fence's longevity. Laying tie-rocks or binders through the fence at regular intervals was characteristic of good rock fence construction and was a technique common in traditional northern English and Scottish fence building (Rainsford-Hannay 1957).

The agreement also provides information about costs and labor. Thornton was to be paid $1.20 for each rod of completed fence. He earned an additional $1.05 per rod if he also quarried and hauled rock to the building site. The records do not specify how many persons assisted Thornton, but contemporary fence construction techniques suggest two additional helpers. They might have quarried and hauled rock and dug the trench for the fence's foundation. Thornton boarded on the Clay farm while he worked on Auvergne's fences. Other notes in the Day Book show that William McGory also boarded on the Stoner Farm in 1844 while he quarried there (Clay 1839-1853).

Other stonemasons also worked at Clay's homeplace. In May of 1848, George Lytle received $52.60 for stone fence work. The price per rod had increased to $3 by this time, so if Lytle were paid at the same rate as the other masons he probably built about 17.5 rods or 288 feet of fence. Later in the year Francis Thornton completed the stone foundation for Clay's stable and, in October, signed another contract to build additional fence:

> Articles of agreement made and agreed to between Brutus J. Clay and Francis Thornton both of the County of Bourbon and state of Kentucky witnesseth. The party of the first part binds him self to pay said Thornton three dollars per rod for quarrying and building stone fence. the said Clay is to have the rock hauled for building the fence and is to furnish the Powder for blowing the rock. and to be at all expences for keeping the tools in order. the said Clay is to board and doe his washing in a Customary maner. and it is understood that each one of the parties may discontinue the work at his option when ever he may Choose to do so—the said Clay is to have the foundation dug out for building said fence. the said Clay is to pasture one horse free of Charge during the Time said Thornton is doeing said work. the fence to be made thirty inches at bottom 18 inches at top four feet high besides the copping. the copping to be well done & all the work in a good workmanlike manner, no work to be paid for untill the fence is up and complete, and then measured, Clay agrees to hire him his boy Melborn at six dollars per month to work with him, but he may be given up at the end of any one month at the option of eather party.
> October 9th 1848
> B.J. Clay
> Frances throonton
> witness J.M. Boyle [Clay and Thornton 1848]

This agreement shows that Clay provided the rock. It also offers the first direct linkage between Irish masons and slaves who worked with them to learn the Irish-Scottish method of fence construction. Thornton's wage for fence work, like that of Lytle, had increased between 1844 and 1848, perhaps reflecting a general increase in prices and wages that accompanied the national economic recovery then under way following the crash of 1837-1842. The wage increase may also have been influenced by local demand for accomplished stonemasons, for this is the period when fence construction proceeded rapidly throughout the Bluegrass. Clay was to provide labor and a wagon or sled to move the rock that Thornton quarried to the fence-building site. Melborn, Brutus Clay's slave, was to work for Thornton. It was customary in the antebellum Bluegrass for large and small slaveholders to hire their slaves to neighbors or town manufacturing firms. Ann Clay, Clay's second wife, kept a list of their slaves, and "Milburn" is one of thirty-six males whose names were so recorded (Clay 1854, June). Clay's mention of board for Thornton and his horse, along with similar notes elsewhere in the record, implies that boarding was a continuing arrangement to be honored when Thornton worked on the farm.

The fence specifications noted in this 1848 document are different from those described in earlier ones. Though their exact location is not noted, they were interior field or lot fences. Later road frontage fences are frequently narrower than interior stock lot enclosures. Since Clay's requirements—thirty-inch thickness at ground level and eighteen inches at the top—are typical of interior lot walls, the fences Thornton agreed to build in 1848 were likely interior ones.

A year later, in November 1849, Francis Thornton received $298 for all rock fence and other stone work finished to that date (Clay 1849). If payment were primarily for fence and quarry work, at the agreed price of $3 per rod, then Thornton would have completed about 100 rods or 1,650 feet of fence. Since records do not specify how much Melborn contributed to the work, it is impossible to estimate how many days Thornton actually spent building fences. Nor is it possible to judge the number of days poor weather prevented fence work.

Clay continued to improve his Stoner Farm property. In November 1849, he credited George Lytle with fifty-seven rods of completed fence and the following June paid him $228 for building fences and "abutments" (Clay 1839-1853; Clay 1850a). These abutments may have been either the heavy stone pillars often placed on either side of a road gateway through a rock fence or the stone piers for a water gap.

Duing the next two years, 1850 and 1851, stone fence construction was a high priority for Clay. A Day Book entry in July 1850, notes that Francis Thornton built additional fencing. He was to receive fifty cents per rod for quarrying his rock, a task for which Clay agreed to provide blasting powder and two laborers (Clay 1839-1853). In addition to Thornton's crew, at least three other crews built fences somewhere on Clay's farm during the same year. Beginning in March, Clay regularly purchased blasting powder and primer from Brian Mitchell, who had a dry goods store in Paris. By September, Clay had bought four kegs of powder and several primers (Mitchell 1850).

Thomas Malone and Michael Conely worked in the quarry and at the fence during the summer and by December 1850 had completed 107 rods and 10½ feet of fence. Malone was paid for about ten rods and Conely the balance (Clay 1839-1853; Clay 1850b). By the following January, Frances Thornton had finished ninty-nine rods of fence on Clay's Green Creek farm, which was apparently a 350-acre tract nearby on Green Creek. About one month later, Thornton began quarrying rock for more fence work at an agreed price of $1.25 per rod, a figure to be doubled if he also quarried his material. Clay also provided a slave, Woodson, to assist (Clay 1839-1853). Later, during the summer, George Lytle and John Burk both completed fence projects on the Stoner Farm, and Clay paid John Gregory for stone work (Clay 1846-1877; Clay Family Papers 1851).

Clay purchased land to add to his acreage. In 1851, D.P. Bedinger wrote to ask if Clay wished to buy land. I have "almost concluded to sell that part of my farm in Bourbon, that lies between your home place and the Paris & Winchester Turnpike Road" (Bedinger 1851). That parcel became the front of Auvergne with a drive leading from the turnpike to the farmstead. Fences along the Paris to North Middletown road, which probably ran along the east side of the property, were also under construction in 1851. Apparently the work crew chose to stack their rock on the road itself, impeding traffic and prompting the following reprimand:

> Sir Take notice that the rock which you have caused to be placed along the line of your fence on the Turnpike Road from Paris to North Middletown & in the gutter thereof, is an obstruction to said Road which not only obstructs the travel on the road but damages the same very much—I am therefore acquired [sic] & directed by the Directory of the road to notify you in writing of the fact & further to state to you that unless the same be removed by you within ten days . . . proceedings will be instituted against you. . . .
> Very Respectfully,
> Jno Jay Anderson Presdt [Anderson 1851]

Francis Thornton again worked on Clay's stone fences during the early months of 1852. Although only a brief note in the Memorandum Book suggests where this work was done, it may be one of the more important citations because it refers to a specific set of fences that stand today. The note reads, "Francis Thorton in full for all work 118 rods [or about 1,950 feet], fence round lot and yard pd cash" (Clay 1846-1877). Several lots adjoin the farmstead to the west with a large orchard lot on the east, and the house is surrounded by a large yard; stone walls separate one lot from another. These fences of excellent workmanship have large, blocky yellowish-brown and gray-brown quarried stones (fig. 5.6).

In addition to identifying an extant fence network, this contract, examined in light of previous agreements, provides a sense of Clay's priorities for fencing. Brutus Clay wanted his pasture and crop land fenced first. Only when that work was well under way did he replace the rail fences demarcating the barn lots and house yard. The records for this year also reveal details concerning the tools employed in fence construction; later, during the summer, Clay had six drills and a crowbar sharpened (Bourbon County Agricultural Society 1852).

From 1853 through 1866, which includes the Civil War period, there is no documented rock fence construction on any of Clay's farms. He commissioned romantic changes to the property about this time—a wrought-iron front porch replaced the classical one, a new stable was built in Gothic Revival style, and Thomas L. Harkins, a "master stone mason," carved lion statues for the tops of the new entrance gate pillars. His brother-in-law, D. Irvine Field, studied at Yale with "Professor Olmsted" (Frederick Law Olmsted) and may have influenced Clay's concern for landscaping and design (Clay Family Papers 1839). Clay had by this time become a community leader and an informed and innovative farmer. The 1850 Federal Census manuscript listed the value of his real property as $80,000. Only one other farmer in Bourbon County had property of greater value. Clay became president of the Bourbon County Agricultural Association and supervised the construction of the fairgrounds (Clay Family Papers 1854-1856). He continued to develop his cattle herd by breeding to imported sires, and he kept detailed genealogical records on his stock. He carried on an active correspondence with farmers throughout Kentucky, Ohio, and Illinois concerning his stock and new farming techniques and ideas. In 1855, a Boston man invited Clay to attend the national meeting of the United States Agricultural Society in Washington and solicited his advice concerning the organization of a stock show at Chicago (King 1855).

By the eve of the war, Brutus Clay's estate had reached impressive proportions. The 1860 Agricultural Census provides a listing of his land and stock holdings: 1,800 acres of improved land with a farm value of $156,000, 30 horses, 10 mules, 40 milk cows, 10 oxen, 400 sheep, and 150 swine. He had harvested or in storage 500 bushels of wheat, 500 of rye, 6,000 of "Indian" corn, 1,000 of oats, and 100 of peas and beans. Together with varying amounts of wool, butter, potatoes, hay, grass seed, honey, and hemp, these were the products of a diversified farm that supplied commodities for family use and a variety of products for the local and regional market.

Brutus Clay was elected to the U.S. House of Representatives and served in Washington during the years of the Civil War. He supported the Union but favored gradual emancipation instead of the immediate abolition

Fig. 5.6 FENCE TYPE BUILT BY THORNTON. This fence section east of the main house is the type built by Francis Thornton. Although he worked with irregular rocks, he succeeded in obtaining close-fitting joints, in part by using the rocks to level the courses. The large coping rocks are typical of all fences on the farm.

of slavery. Clay's was a unique position in the legislature. During the war, one of his sons was an officer in the Union Army and another in the Confederate. Because of his sons' influence, neither army seriously damaged the Clay plantation when they each occupied Bourbon County, although the troops took much of his prized stock for provisions. His wife, Ann, and daughter, Martha, competently managed the farming operations while he was gone; the farm continued to function at a profit, thereby preserving the Clay fortune (Clay 1990).

In the spring of 1867, rock fence building began again. Although the slaves were emancipated during the war, some continued to work for Clay as freedmen. A note in the Day Book in April 1867 reads, "Joe commenced building stone fence on the road he worked 4½ days in the quarry before starting to build fence." A "Joe" is listed in Ann Clay's Negroe Book of 1854. If this is the same man, we may assume that Joe had worked as a mason or mason's assistant before and not only knew how to quarry stone but could build fences. Quite possibly, Joe was one of the men that Clay assigned to help Thornton or the other masons before the war. Less than two weeks after this entry, another lists Bill Kennedy as building fence with Joe (Clay

Fig. 5.7 ROAD FRONTAGE FENCE. This fence was probably built by Bill Kennedy and "Joe." It is not as tall as the lot and pasture fences. Workmanship here is clearly different from lot fences; many joints between courses are not fully covered. The masons did, however, build a full-width cap course and use large coping rocks.

1854-1875). By the last of September, the two men had completed seventy-nine rods of fence along a road. This could have been the road that fronts the farm and runs from Paris to Winchester (fig. 5.7). A few days later, Joe began a fence that was to run across Kennedy Creek near Clay's neighbor Jesse Kennedy. Bill Kennedy also worked on the fence. Together, he and Joe built some seventy-four rods, including the piers where the fence crossed the creek. Joe received $84 for his work; Bill Kennedy received $50 for his (Clay 1854-1875). The fact that Clay paid Bill Kennedy for working on the fence implies that these fences were not property boundaries between Clay and Jesse Kennedy; otherwise, Kennedy would have been expected to share the building cost. Bill Kennedy, the stonemason, was not necessarily associated with the neighbor of the same surname, although he may have been a freed Kennedy slave.

More fence building took place in 1868, and in February 1869, Clay measured those built by Robert Griffin and "King," which ran "from the gate to Kennedys Creek." The original gate opened to the old lane which ran from the northwest corner of the pasture to Kennedy Creek. Clay measured the fence from there to the creek as 51 rods, or about 840 feet. He paid the masons $140.25, or $2.75 per rod (Clay 1854-1875). This is one of the few fences on the farm whose builders are explicitly identified in the record (fig. 5.8).

Two years earlier, Joe and Bill Kennedy had received $2 per rod. Since the national economy was in a post-war boom, this wage increase could have represented inflation. Alternatively, Clay may have acknowledged differences in masonry skills and paid according to the quality of the finished fence. Joe and Bill Kennedy apparently quarried and hauled their own rock. Although the Day Book entry does not record whether or not Griffin and King quarried their rock, they likely did, for the wage they received was uncharacteristically high for the period.

In mid-February 1869, Irishman John Dowd joined Griffin, and the two men "commenced stonework and Bull lots." They received $4 per week instead of payment by the rod. Just over one year later, Dowd received $116.85 for "all stone work done in building stone fences" (Clay 1854-1875). If John Dowd was paid $2.75 per rod, this amount would have been for some 42 rods, or about 700 feet. The precise location of Dowd's fences has not been identified, but the bull lots in the west section of the farmstead contain over 200 rods of stone fence, and foundation stones still in place imply that at least

Fig. 5.8 GRIFFIN AND KING FENCE. Kennedy Creek runs behind the fence built by Robert Griffin and "King." Tie-rocks in this section extend three to four inches from the face on the opposite side. Ruins of the Thomas Kennedy stone house stand in the background.

Fig. 5.9 BULL LOT FENCES. Southwest of the Clay house, livestock lots of an acre or less are surrounded by rock fences.

50 rods more had been removed at a later date (fig. 5.9). These fences would have accommodated Brutus Clay's expanding interests in stock breeding. His bloodstock genealogical records and the letters of inquiry about his bulls and cows indicate that he was a highly regarded breeder. The fenced lot complex would have allowed him to pen stock selectively, giving his bloodlines credibility.

Only limited stonemasonry is recorded in the Day Book or Memorandum Book after 1870. In 1873, John Dowd and John Doyle were paid, but for what work we are not told. In August of that year, D.C. Clay (who may have been a freedman), was paid $50 for stone work (Clay 1846-1877). Notes in the Day Book and Memorandum Book through the years show that construction of new fence and repair of existing fences occurred simultaneously, usually by the same men.

Drawing Conclusions

Although Brutus Clay's estate was larger and more valuable than those of most of his neighbors, the record contains no evidence that either the crops that he grew or his mix of stock differed substantially from other farmers in Bourbon County. He specialized in producing cattle that were in demand as foundation breeding stock by farmers in the Ohio Valley. Clay recognized the unique qualities of the Bluegrass as a place to breed quality stock, as did his admirers and colleagues. Noting the success of earlier cattle importations and the quality of their progeny, a Lexington cattleman, John Cuse, wrote to Clay in June of 1855, commending Clay's essay in the *Ohio Farmer* that spoke to the superior qualities of Kentucky-bred cattle. Cuse thought that while Ohio farmers must have been frustrated by competition from the South, they grudgingly recognized that "Kentucky [was] the England of America" because farmers there produced superior stock through superior circumstances.

Clay's wealth was not based solely on his Bourbon County farm; he had extensive landholdings elsewhere in the state and in Missouri, Illinois, and the Mississippi Delta. Nothing in his records, however, suggests that he built substantially more rock fencing than other farmers in the Inner Bluegrass region. In fact, in southeastern Bourbon County, in the steep hills of the Eden Shale area, farmers of much more modest means built comparable rock fence networks on small and medium-sized farms. Fences on Eden Shale farms are often of field and creek rock. Clay's fences are of quarried stone as were those of most farmers on Maury-McAfee soils. While Clay's operation was substantially larger than the farms of many of his neighbors, his actions and decisions were similar to theirs.

Brutus Clay's records portray the use of professional stonemasons for rock fence construction. Although Clay was a major slaveholder, he clearly preferred experienced fence builders. The Irishmen who worked for Clay brought a practical skill of their homeland, a craft in demand in the Bluegrass. They provided an increasing supply of skilled laborers who worked for board and low pay building fences. Almost all of Clay's fences were built on a piecework basis instead of for a wage.

Clay employed twenty-four different quarriers and masons between 1836 and 1876, many of whom were probably itinerant workers. The census schedules from 1850 to 1880 substantiate that at least thirteen of these men were Irish. Their names are Jonathan Boone, John or Jonathan Burk, John Carty, Michael Connelly or Connell, John Dowd, John Doyle, John Gregory, Thomas Malone or Malin, Patrick McGory or Gory, William McGory, James Moloy or Malona, Francis Thornton, and Page or Pagy Ward. Many Irish emigrants who fled their country during the famine years were very poor, and many were illiterate. The contrast between Brutus Clay, who was educated in the Greek and Latin classics, and Francis Thornton, an important Irish mason who signed his construction contract with an X, is striking.

Clay did not explain his choice of masons or elaborate upon their skills or shortcomings. Existing fences on Auvergne demonstrate that different masons with different skill levels constructed them. The individual stones in some fences are so carefully positioned that even after they have stood for well over a century, one cannot insert a knife blade between them, and all

Fig. 5.10 WELL-BUILT FENCE. Rocks in this Auvergne fence are carefully fitted and most joints are covered.

Fig. 5.11 FENCE WITH RUNNING JOINTS. This fence exhibits less skill than many other of Auvergne's fences. It has running joints and very irregular coursing.

joints are carefully covered (fig. 5.10). Other fences show less care and have frequent running joints (fig. 5.11). All of them, however, have battered faces, foundation and tie-rocks, cap courses, and coping stones. In structure and appearence they are like other fences built across the region. Further, they are exactly the same as Scottish, Irish, and northern English fences.

If Kentucky slaves had originated fence building, their fence design could not have coincidently been the same as the fences of the British Isles. How they came to be involved in rock fence building is illustrated by the records of Auvergne. Slaves participated in stone quarrying and fence construction as helpers or assistants to the Irish masons but were not lead masons on fence projects until after the war. The Day and Memorandum Books, other notes, and letters frequently mention slaves' work and health. If they had built fences, the fact undoubtedly would have been recorded.

Brutus Clay did not make explicit his rationale for building stone fences. His records lend depth to our understanding of how the fences came to be built and document the link with Irish masons, but nothing in them implies that his stone fences were a measure of status nor that they were built to improve the appearance of the plantation. The fences served the specific functions of confining stock and demarcating property boundaries; availability of building material and permanence were primary concerns.

CHAPTER 6

Change and Legacy

> The past as we know it is partly a product of the present; we continually reshape memory, rewrite history, refashion relects.
>
> —David Lowenthal, *The Past Is a Foreign Country*

The end of the Civil War brought less change in Kentucky than in the Deep South. After emancipation, some freedmen left their owners' farms and moved into towns and cities seeking work. Others became day laborers or tenants and lived in hamlets created to house former slaves or in tenant houses on the farms of their new employers (Channing 1977, 136-39).

The farm economy began to change in the Inner Bluegrass, but new building forms and methods of subdividing farms appeared slowly. Large farms were not divided into sharecropper units as in the Cotton Belt (Prunty 1955). Livestock for draft, for breeding, and for meat continued to provide handsome profits on Bluegrass farms and estates. Cattle importations from England resumed, and the domestic shorthorn breeding market flourished. Before the war, the region had been an important horse breeding area and a supplier of mules to the plantations of the old Cotton Belt (Hilliard 1984, 45). The horse industry suffered serious thefts by northern and southern cavalries and local guerrillas who stole select breeding and racing stock for military service or ransom (Clark 1968, 177-78). The infusion of outside money, however, aided economic recovery.

The great financial panic of 1873 played havoc with the region's farms. Small farmers could not meet mortgages and either lost their land and moved on or stayed to become tenants. Heirs divided farm acreage or sold it to cover debt. The number of farms increased and the average farm size decreased. Central Kentucky farmers gradually came to be either smallholders or operators of large estates, with relatively few in the middle ground (Channing 1977, 157-58). Labor costs rose. Freedmen might receive eight to ten dollars per month as farm laborers, whites a few dollars more. Irish immigrant laborers also received higher pay.

People made substantial adjustments in the way they organized their land and farms, including fencing their fields. Statements in the agricultural press indicate that rock fences built pior to the Civil War not only functioned as basic enclosures but also came to be measures of a well-managed farm. After the war, rock fences were increasingly expensive and were built only on large estates. Gradually rock fences became symbols of elite social and economic status. Although stone was readily available at the cost of quarrying, a fence mason's labor was dear. More seriously, few masons remained who were trained in fence construction using the traditional Scottish methods that produced superior fences, soundly built with foundation stones, tie- or binder stones, cap courses, and large triangular coping stones. Only wealthy gentlemen farmers could afford to build great lengths of rock fence (cf. Raup 1947, 7).

Some large, well-managed estates recovered from the war sufficiently so that by the 1870s they were again producing cattle as breeding stock for western markets, primarily the southern Great Plains. When members of the English Agricultural Interests Commission visited the Bluegrass in 1879, they found a number of large shorthorn breeding farms with first-class stock. Among the farms that they visited was Woodburn, in Woodford County. The British commissioners found, in 1879, a "grand herd of Bate's blood" (a strain of shorthorns). They tabulated "50 head of beef cattle, 60 head of Durhams, 40 of Jerseys, and grades. Of blood horses, 40 yearling thorough-breds are sold as mares at 300 to 400 dollars each" (Harrison 1970, 274). Annual shorthorn sales from the farm averaged between $75,000 and $130,000.

Robert Alexander, a Cambridge-educated Scot, had established this august place in the 1790s. His nephew, Robert Aitcheson Alexander, added thoroughbred and trotting horse breeding to the farm's activities and by 1856 was producing winners at the track (Clark 1968, 176-77). Robert Aitcheson Alexander also commissioned a long rock fence built around the perimeter of the farm (*Valley Farmer* 1857). By 1879, however, fencing preferences on the farm had changed. The cattle were pastured in enclosures "made with black walnut sawn rails 1½ inches thick by 6 inches wide, fastened to split locust posts; these [would] last 30 years, and the rails 15" (Harrison 1970, 274, 280). This brief report indicates that another change in the mode of farm fencing was under way. Unlike Brutus Clay, R.A. Alexander built rock boundary fences but no interior rock fences. The sawn plank and post fences described by the Englishmen were to become the quintessential Bluegrass livestock farm enclosures and symbols of the gentleman's rural retreat, though oak planking would replace walnut.

In the 1890s, John Burroughs found the Bluegrass as pleasing as England's farmlands, commenting that in Kentucky the fields were larger and "not so cut up, nor the roadways so narrow, nor the fences so prohibitory" (Burroughs 1895, 227). The fields were often fifty to one hundred acres, giving the appearance of openness and requiring less fencing.

Kentucky writer James Lane Allen did not see the countryside as the open expanse Burroughs described. The Bluegrass landscape had been undergoing change since the 1870s as the wealth of the estate owners increased. The display of large homes, great barns built of fine hardwoods, landscaped grounds, graded training tracks, and plank fences had substantial visual impact. These farms of industrial tycoons dominated only small sections of the select limestone lands. Local people, some of whom had been landowners for generations, made their livings from the remainder. Allen thought that Kentuckians retained their ancestors' love of enclosures. He wrote, "One does . . . notice here and there throughout the country stonewalls of blue limestone, that give an aspect of substantial repose and comfortable firmness to the scenery. But the farmer dreads their costliness, even though his own hill-sides furnish him an abudant quarry. Therefore one hears of fewer limestone fences of late years, some being torn down and superseded by plank fences or post-and-rail fences, or by the newer barbed-wire fence" (Allen 1892, 26-27). As Allen confirms, rock fences were out of favor, largely because of high labor costs.

Roads vs. Rock Fences

The symbiosis between road and fence, turnpiker and mason, was eroded as private turnpike construction companies surrendered control of roads to the state. A revised road statute, Sims Road Law, passed in 1894 enforcing regulations concerning proper engineering techniques (Allen 1954, 252-57). The law specified that dimensional stone be used in retaining walls and culverts and that the width of the right-of-way be strictly enforced. This meant that "all fences, buildings, and wood on the line of the road, if not removed within a reasonable length of time after notice is given shall be cleared off by the contractor, piled in such a manner as the supervisor or contractor may direct, and preserved for the use of the owner, but at his expense" (Crump 1895, 14) Rock fences stood at the edge of central Kentucky roads for three-quarters of a century. Now they were to be removed, or set back if the landowner wished and could afford to hire the masons to rebuild the fence. More often landowners donated the fence to the road-building effort, and turnpikers, using round-headed stone hammers on long springy handles, reduced the individual stones to two-and-one-half-inch cubes to be incorporated into the road bed.

Fence removal became even simpler, as did road building in general, with the arrival of the steam-powered rock crusher (fig. 6.1). These primitive-looking devices consisted of a rotary rock breaker, a steam engine and boiler, a revolving sorting screen, and bins to hold the separated product. The entire assembly could be towed about by a team of mules. The large crews once needed to break rock by hand dwindled to a few men needed to haul it to the machine, keep it full and functioning, and return the crushed pieces to the road bed.[1]

Fig. 6.1 STEAM ROCK CRUSHER. During the depression, WPA laborers used rock crushers to break rocks taken from fences, fields, and quarries into small aggregate for road surfacing or into fine particles that were spread on fields as agriculture lime (1936 photograph courtesy of M.I. King Library, University of Kentucky, Goodman-Paxton Collection).

A Boyle County landowner recalls the era: some road-building material came from large roadside quarries using a Norville steam shovel and mule-drawn wagons, but many roadside rock fences were taken down during the 1920s and the rock used in road construction (Caldwell 1989). When farmers hauled stone, including relict rock fences, to the roadside from their land, they received compensation for what they provided (Fletcher 1907, 10-11). The crushers proved handy machines, and farmers and road crews used them into the 1930s to break up rock fences.

Rock Fence Demise

Between the turn of the twentieth century and the end of World War II, farmers changed to a more commercial orientation. The gasoline-powered tractor gradually replaced draft horses and mules. Sheep-raising diminished because of problems with wild dogs and parasites. Beef cattle increased in number, and dairy cattle became a specialty of some farms close to the larger metropolitan markets in Louisville and Cincinnati. Farmers grew burley tobacco in such prodigious amounts that in the 1930s the federal government created an allotment program to stabilize prices. Corn, wheat, and oats were grown as livestock feed or sold for milling. Enough farms specialized in stock breeding that the *Kentucky Farmers Home Journal* resurrected the idea that the state was still "the England of America" for purebred livestock (1941).

Beyond the Inner Bluegrass, in the Eden Shale hills, change came less rapidly. Farmers had cultivated the steep hillsides long enough to learn that they would harvest mostly rock and more rock. They allowed the eroded hills to revert to woodland or planted fescues, and people retreated to valley bottom tobacco patches or moved away. Such poor land did not warrant investment in new buildings and other improvements, and the region seemed destined to remain a repository for a large collection of old log houses and outbuildings.

At the same time, Inner Bluegrass farmers adjusted their farms to accommodate the larger, more efficient fields that the tractor encouraged. Adjustment meant removing interior fences so that smaller fields could be combined. Where those fences were stone, farmers often employed the rock crusher to reduce fence rocks to granules, which they spread on the fields to enhance fertility. Many changes on central Kentucky farms were not simply the results of market forces or new technology, however. The federal government, through Department of Agriculture publications, urged improvement, specialization, and conservation. Furthermore, the private agricultural press was still active, as it had been for a century, urging "better practices" by the farmer (Fusonie 1977, 33-56).

Agricultural journalists seemed particularly intolerant of ineffective practices and unwarranted costs. They pointed out that rock fences were a major source of expense and inefficiency. Since an adequate enclosure might cost one dollar to build for each dollar's worth of livestock it restrained, building and maintenance costs were foremost concerns (Roberts 1905, 336). The invention of a method for storing corn and other fodder crops in a silo solved part of the fencing problem (Noble 1981, 11-14). This innovation allowed the farmer to stable stock during cold winter weather or the hot months of late summer, requiring fewer pastures. At least one farm journalist advocated eliminating half of the fences on the typical livestock and crop farm, keeping only those that surrounded permanent pasture land (Roberts 1905, 338-39). This kind of consolidation destroyed the interior fences that divided many farms into small lots.

After the creation of the U.S. Department of Agriculture and the Land Grant College system in 1862, new ideas in farming were increasingly based on scientific research and published for the general agrarian readership (Pinkett 1977, 159-68). Urging farmers to conserve soil and improve management, this new generation of agricultural writers largely ignored the aesthetic value or historic merit rock fences might have. They instead told farmers that rock fences presented manifold problems. Frost heaving might cause improperly built fences to "creel" (or fall out), requiring repair by a mason. Rock fences harbored brush and weeds and provided homes for insects or burrowing animals. Further, to rebuild those in need of extensive repair was impractical because of labor cost (Humphrey 1916, 4, 10).

Agricultural writers further claimed that the well-built wood or wire fence was good insurance against lawsuits and promoted good relations between neighbors. In addition, the neat wood or wire fence added to the farm's attractiveness and declared to passersby one's concern for thrift and appropriate management (Kelley 1940, 1). If farmers did not see the folly in rock fences, the agricultural press was quick to point it out to them (Fox 1958; Fox 1959). Barbed and woven wire were widely available by the 1890s and became cheap and practical replacements.

A formidable combination of the law, science and technology, changing agricultural production, and alternative fencing materials acted in concert to clear the land of most of its rock fences. By 1904, a popular picture book of Bluegrass estates included a photograph of a painstakingly constructed rock fence and the author's comment that it was "a type of fence that is fast disappearing" (Knight and Greene 1904, 12).

Rock fence destruction continued into the depression, especially in the Eden Shale hills. The Civilian Conservation Corps (CCC) established an office and camp for men at Carlisle in Nicholas County, and the U.S. Department of Agriculture and the Soil Conservation Service coordinated crews to perform work projects. The work involved reclaiming eroded farmland, clearing overgrown pastures, improving soil fertility, and generally upgrading areas where the land had been abused by unwise cultivation. In the shale section of Bourbon County, near Little Rock, and in adjoining Nicholas County, CCC crews gathered rock from hilly fields and pastures and stacked it into large oval or circular piles (fig. 6.2). Similarly, Rollinson

Fig. 6.2 CCC ROCK PILE. This coursed circular mound of field rock, gathered by CCC crews in the 1930s, is one of many still standing in the Eden Shale section of Bourbon County. The foreground fence is field rock.

([1969] 1972, 16) reports that clearing the land of surplus stones also included the piling up of "clearance cairns" in the middle of fields in Great Britain. While a few stone stacks remain in the Bluegrass as striking reminders of the scale of work done, most were hauled to portable rock crushers and ground into agricultural lime, which workmen spread by wheelbarrows onto surrounding fields. Rock fences also fed the crushers, enlarging fields and eliminating the need for costly fence repairs (Clepper 1973, 11).

Within a few years, however, the region's most perceptive landscape observers began to realize that such extensive fence removal was a serious error. Clem J. O'Connor, writing for the *Louisville Courier-Journal* in 1952, believed that the fenced landscape possessed unique character. "In this day of inflationary costs," he wrote, "it would be well-nigh impossible to evaluate these fences. . . . A few years back, quite a number of them were taken down and replaced by woven-wire fences. The limestone was ground and spread as fertilizer on the fields. This was, of course, a mistake, for their value as a distinguishing characteristic in a state seeking tourist trade far outweighs their value as agricultural limestone to the fields on which they are spread" (7).

Fences as Landscaping Features

If working farmers sometimes pulled down their rock fences to expand fields or avoid expensive repairs, the owners of the large Inner Bluegrass estates revived rock fence building around the turn of the twentieth century. During the 1870s and 1880s, wealthy outsiders bought large acreages of limestone land and began breeding fine blooded cattle and racehorses. A few natives, having made their fortunes elsewhere, returned to buy and expand existing farms. This substantial infusion of new money had a dramatic impact on the scale of Bluegrass agriculture and on the traditional economy. These individuals also brought an outside perspective that affected people's values and tastes with a new estimation of the good life. The new aesthetic included an idealized landscape that came to be symbolized by manicured pastures, great houses, and horse barns (Clark 1968, 176-83) (fig. 6.3).

North of Lexington along the Newtown Pike, for example, Price McGrath's 2,200-acre farm, McGrathiana, passed to thoroughbred horse breeder Milton Young in 1882. After the turn of the century, the C.B. Schaffer family acquired this land, built new buildings, and renamed the place Coldstream. A quarry on the farm furnished tons of rock, which masons used to build barn foundations and lot walls. Between 1915 and 1920, Schaffer had three miles of rock fence built along the pike fronting the farm (fig. 6.4). Masons set ashlar stone with concrete at the crescent-shaped drive entrance and quarried stone in the remainder of the fence. The coping stones are small, upright, and sealed with a concrete top. George Smith, a black

Fig. 6.3 HORSE FARM FENCE. The mortared fence fronting a lane in Fayette County is of quarried rock laid in a random ashlar pattern. The horse barn displays features that have come to epitomize Kentucky horse barns: Palladian windows, cupolas, and cross-hatched doors.

Fig. 6.4 COLDSTREAM FARM FENCE. A quarried rock fence on Coldstream Farm in Fayette County separates the entrance drive and house from a woodland pasture. The fence has small coping rocks set vertically and surfaced with a thin coat of cement.

man from Georgetown, in Scott County, directed the construction crew that built the fence (Hall 1989).

One of the first to reshape this land in a substantial manner was James Ben Ali Haggin, who, at the turn of the century, purchased property along Elkhorn Creek north of Lexington, eventually accumulating some ten thousand acres. Born in central Kentucky to Ulster-descended pioneer stock, he was one of the largest landholders in Kentucky at the time. Like many of his wealthy contemporaries, Haggin accumulated his fortune not in his home state but in California, where he practiced as an attorney. He parlayed profits from irrigation water disputes into California farmland, as well as Montana and Chilean copper (*New York Times* 1914, 9; *San Francisco Examiner* 1914, 1). He called his Bluegrass property Elmendorf Farm and had architects erect new barns and innovative outbuildings, including a large dairy and an electrical powerhouse. Construction crews built or improved roads to serve the estate (*Louisville Courier-Journal* 1914, 1). Masons erected an elaborate cut stone entrance along the Ironworks Pike and another that incorporated a small stone waiting room along the trolley car tracks that paralleled the Paris Pike (fig. 6.5). Haggin's formal stone farm entrances were an imitation of those on large English estates, and they soon became de rigueur for Bluegrass farms (Rackham 1986, 129-39).

Fig. 6.5 ELMENDORF ENTRANCE AND WAITING ROOM. A formal rock entranceway at Elmendorf Farm in Fayette County once faced the Interurban Railroad between Paris and Lexington. The low ashlar fence on the left is topped by a flat coping; beyond the left corner post the coping is castellated.

When Haggin died in 1914, his estate was divided into smaller units. Joseph E. Widener, a Philadelphia businessman and horse fancier, purchased one part. Widener's architect, Horace Trumbauer, designed a new set of buildings, including a new training barn for which he "chose the English court yard plan" that featured two rows of twenty-four boxes or stalls, each facing a central court. The structure was roofed over and surrounded by a covered one-sixth mile training track (Widener 1940, 231). Though reduced to 1,700 acres, Elmendorf Farm underwent a positive transformation. Extolling the virtues of the farm, the *Thoroughbred Record* recounted the dimensions of the changeover: "Improvements include the building of . . . model training and yearling stables, remodeling of a beautiful century-old house; the restoration of old slave quarters, the construction of twin bridges over the Elkhorn and a tributary . . . the planting of thousands of trees and shrubs, [and] installation of a lake and game preserve" (1926, 318).

Widener's crews moved more than one hundred thousand yards of rock and earth. Clearly, great wealth could affect great change on this land, and any landscape feature thought important for function or for fancy could be built. Even stone fences were not too expensive for Widener. During the depression's early years in 1931-32, he commissioned black stonemasons to build a stone wall along the front of the farm on the west side of the Paris Pike (fig. 6.6). These masons had apprenticed with Irishman Edmund P. Woods (Miller 1989a).

Fig. 6.6 ELMENDORF FENCE. The rock fence north of Elkhorn Creek along Paris Pike was built during the Great Depression by black stonemasons when this acreage was part of Elmendorf Farm. Coping rocks are carefully sized and shaped and stand together like books on a shelf. Some fence sections contain mortar.

On another portion of Haggin's Elmendorf Farm, James Cox Brady, a New York traction magnate, spent substantial monies to make over Dixiana Farm. Between 1926 and 1928, Brady's workmen planted trees, built new buildings, erected new wooden fences, and built stone bridges across Elkhorn Creek (fig. 6.7) and a large, formal stone entry (Jordan 1940, 56-60).

Many other gentlemen farmers commissioned new stone entrance gates and road frontage fences. Masons often constructed this new stone work in such a way that it bore a strong resemblance to the traditional Scottish- and Irish-type fences erected before the Civil War. Yet the type of rock used and the fences' internal structure were often different from those of the old fences, in part because few remaining masons had the knowledge and training to build traditional dry-laid structures, but also because the mason's concern for proper technique was sometimes subsumed by the landowner's concern for ostentation. New walls were often built of stone quarried dozens of miles away rather than from beds that lay a few feet below the turf and were often bonded not by gravity in the Old World tradition but by concrete. That these new frontage fences were to serve primarily as decoration became clear when farm owners built post and plank fences a few feet inside them (fig. 6.8). The wooden enclosure, then, contained the valued stock, while the stone structure became a subaltern whose function was to be seen by passing motorists and to lend focus to a large formal entrance crescent.

While black fence builders formed the major work force constructing rock fences bordering horse farms in the twentieth century, the Irish also continued the practice. Simon Holleran, born in 1837 in County Clare, Ireland, migrated to Kentucky in 1854. He built fences for the Hancock family on the road frontage of Claiborne farm in Bourbon County in the early 1900s (Holleran 1990). He and his brothers also built more than 300 miles of road in Bourbon and Montgomery counties. In 1953, James Rogers Gormley, son of an Irishman, built rock fences on the Fayette County horse farm of C.V. Whitney (Gormley 1987). Some fence builders today are descendants of Irish stonemasons (Waugh 1988).

Reading the Contemporary Landscape

Examples of old rock fences are scattered across the Bluegrass. The plantation-era fences that created internal divisions on farms were the first to go when agricultural specialization began after the Civil War. Although some farm boundaries are still marked by rock fence remnants, interior lot enclosures are rare and hard to find. Those that remain are often on family farms in which each generation, or each succeeding owner, has recognized the fences' aesthetic and historical merits and has had or found the wherewithal to keep them in good repair. Crop and livestock management may

Fig. 6.7 STONE BRIDGE AT DIXIANA. The bridge across North Elkhorn Creek at Dixiana Farm in Fayette County features mortar-bonded masonry similar to the entry gate crescent at the farm entrance.

Fig. 6.8 PLANK FENCE INSIDE ROCK FENCE. The black plank fence stands inside the rock fence at this Fayette County horse farm, protecting valuable horses from visitors and from being scratched by the rock fence. Such fences are common along horse farm road frontages.

have changed on these farms, as on others, but priority lay with fitting a new system into the network of rock fences rather than obliterating them to erect plank or woven wire.

The greatest number of old fences remain in the Eden Shale hills. In Owen County, for example, miles of edge fences built of massive stone blocks run through the woods. Dozens of miles of fence still stand in western Anderson and Mercer counties (fig. 6.9). Because most of these fences were built in edge coursing and were carefully constructed, they have as much integrity today as they did the day they were finished. In Mercer County, many hillside pastures are cleared and the fences are visible from the roadside. Runoff control walls and rock piles from field clearance are also numerous there. In Anderson County, much of the shale land has reverted to trees, and woodland obscures the fences that remain (Davis 1927, 48-49). These fences, like the old log houses and barns they surround, exist simply because no one needed to remove them. Farming in the Eden Shale hills is not very profitable and continues only on a modest scale. Landowners see little need to clear the eroded hills of trees again or to gather field rock. The fences do not stand in the way of improvements, and they remain as signatures of men of another era who were part of a process that has run its course.

Fig. 6.9 FENCE ROWS ON AN EDEN SHALE HILLSIDE. Field rock fences in edge coursing traverse this western Mercer County land, and cedar trees are beginning to colonize the fields in the distance.

Fig. 6.10 KENTUCKY RIVER VALLEY FENCE. A field rock fence ascends steeply sloping land north of Shakertown's West Family House, a few miles from the Kentucky River. The excellent condition of this horizontally coursed fence, built of irregular rock on rough land, is testimony to the skill of its builders.

The most rugged land of the Inner Bluegrass is the steep creek valleys that drop into the Kentucky River gorge. There, as in the Eden Shale, plantation-era fences survive. These rock fences often enclose old grist and lumber mill sites, while others border roads. Still others climb steep slopes to mark the boundaries of old fields (fig. 6.10). Stone from the fields and small ledge quarries went into these fences near the gorge. In the deeper valleys, fences built of the blocky white Tyrone rock remain. Few large farms occupy these steep-sided valleys where the land is best used as pasture or not used at all. The small farms on the perimeter of the Kentucky River gorge incorporate the old fences into their enclosure network with little need to do more than keep them in good repair.

Inner Bluegrass turnpike fences built in the middle to late nineteenth century stand in significant numbers on road frontages. Although hundreds of miles of fencing built by Irish and free black fence builders are now gone, those that remain are handsome examples of the stonemasons' art. These fences are sometimes fragments that may not connect to other fences or fully enclose a field or pasture. Few roads have rock fences aligning both sides as they once did. Inner Bluegrass counties have only a few paired turnpike fences standing to illustrate the landscape of a century ago (fig. 6.11).

Fig. 6.11 PAIRED TURNPIKE FENCES. As recently as the 1920s, many roads in the Inner Bluegrass were lined on both sides with rock fences, as is this land bordering horse farms north of Lexington. The fence in the foreground has extended tie-rocks, an extended cap course, and heavy coping rocks, following early nineteenth-century Scottish traditions. The fence on the far side of the road has recently been rebuilt.

The Fence Environment and Damage Potential

Many property owners expend considerable effort to maintain surviving fences. Even the highest quality fences are subject to damage from a variety of sources. The annual freeze and thaw cycle poses a significant threat. If the ground freezes to a depth of six inches, a sunny winter day may cause the ground to thaw on the south side of a fence while the north side stays frozen. The south side may settle and sag, and eventually fall-out may occur. Even if the fence has a deep foundation, the rocks on its south face may freeze and thaw in a daily cycle, causing it to weather more rapidly than the north face. Fences are particularly vulnerable during the first warm days of late winter. As temperatures rise and frost leaves the ground, the topsoil turns to the consistency of pudding. A heavy rain will aggravate such instability. A well-built fence should stand, but the vibration of a passing truck or a stout thunderclap can cause portions of a poorly constructed fence to creel.

Rock fences have offered generations of birds convenient places to perch, but the price has been high. Tree seeds contained in droppings land beside the fence, beyond the reach of browsing stock and the sweep of scythe and sickle. As the young sapling grows, its roots extend into the cool dampness beneath the fence, and the trunk grows upward, encroaching on the fence face. Eventually, the trunk may push the fence out of alignment (fig. 6.12). Well-built fences may shift a foot or more and tilt away from the trunk fifteen degrees before gravity overcomes the binding power of the tie-rocks and cap course so that the fence falls. Tree limbs may break off in heavy winds and land on the fence below, causing damage, but the well-built fence can resist even this assault: both horizontally coursed and edge fences stand without noticeable damage after limbs a foot in thickness have fallen and broken across their copings (fig. 6.13). Small saplings, grapevines, grasses,

Fig. 6.12 TREE PUSHING FENCE. This quarried rock fence in Woodford County was probably built in the 1850s. As trees grow along the fence's flanks, the trunks and roots gradually push the fence out of alignment and will eventually cause it to fall out.

Fig. 6.13 TREE FALLEN ON FENCE. Expertly constructed fences like this one in Bourbon County can withstand blows from falling trees and branches with little damage. Through rocks tie the fence together and help resist movement. This fence is one of a former pair that bordered a now-abandoned road.

and weeds grow up along rock fences, hiding them from view and sheltering damp mosses that may cause deterioration. Groundhogs seem to prefer such coverts and their tunnels can be as costly to fence integrity as roots, frost, and thunder.

Modern Removal

During the 1960s and 1970s, Lexington was among the twenty most rapidly growing cities in the United States. Rapid suburbanization extended a broad swath of streets and tract housing across the Bluegrass countryside. Other county seat towns in the Inner Bluegrass, such as Danville, Versailles, and Georgetown, sprouted subdivisions. City expansion has not abated; houses on small lots now line country lanes as farm owners sell off road frontage. The city's encroachment onto rural land has had a dual impact on rock fences. Some fences are removed simply to make way for new buildings. One shopping center contractor in need of parking lot space razed a quarter mile of old rock fence with a bulldozer and buried it in one afternoon. When roads are widened, as they have been to connect Lexington with surrounding areas more efficiently, the fences that bordered them are buried or hauled away.

The suburb also claims fences in a more subtle way. For forty years or more it has been popular to veneer houses with fence rock. One means of dismantling fences to obtain rock for veneer is for a contractor to buy a fence from a farmer who cannot afford to maintain it. The price offered will allow the farmer to replace the fence with new woven wire or plank. The contractor gains material that, when sized and fitted, presents a handsome weathered facade. For some homeowners, it is apparently a matter of pride to tell guests (erroneously) that their house was faced with many rods of "slave-built rock fences" (Eads 1953, 34-35).

The city is not alone in consuming old rock fences. Some large country estates that have no standing fences are buying them in outlying counties and trucking the rock to the property for rebuilding (fig. 6.14). The process can be painstaking and expensive. Some work crews begin by laying a bed of sand in the bottom of a large dump truck. Others may use a flatbed wagon. The fence is taken apart, rock by rock, and laid into the truck bed. When loaded, the truck is driven to the building site where it is carefully unloaded onto a stockpile for use in erecting new fences. Much rock is discarded because it is too weathered or damaged. Many progressive fence masons, however, do not use old material but creek rock or new stone quarried for the purpose.

Estimating the amount of rock fencing that once stood in the Bluegrass is difficult. Anecdotal information from people who have lived in the region for decades suggests that only 5-10 percent of what was one of the most extensive networks of quarried rock fences on this continent remains. A rock fence mason who works in Bourbon, Fayette, and Scott counties, for example,

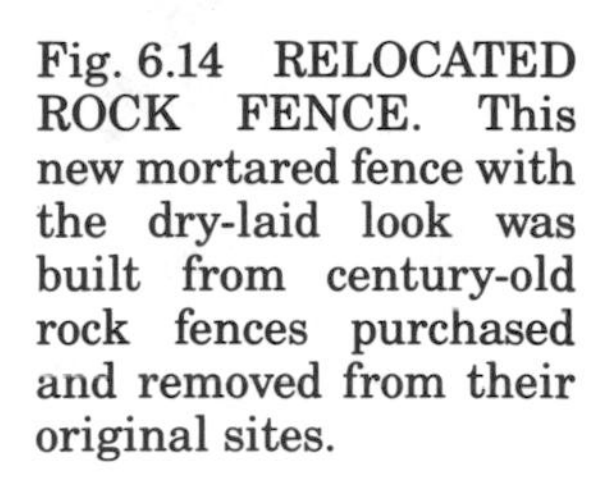

Fig. 6.14 RELOCATED ROCK FENCE. This new mortared fence with the dry-laid look was built from century-old rock fences purchased and removed from their original sites.

can remember when almost all the roads in these three counties were bordered on both sides with rock fences (Waugh 1988). With so many factors conspiring to remove fences or make them expensive or difficult to repair, the remainder are vulnerable to continued encroachment. Urban growth and change are subsidized at the expense of heritage and scenery.

Locating razed fence networks is possible. They seem to survive in proportion to the skill and dedication with which they were built. Large rock abutments often remain on the banks of creeks that fences once spanned. Deep-set foundation rocks are left behind, revealing themselves as a line of sparse grass across a closely grazed pasture where a fence stood. Probing the ground with a metal rod or spade may reveal rocks along lanes covered by tall grass or weeds. The spalls from the heart of the fence leave a trace on the ground where they spilled during removal. Easiest to read are "fence shadows"—trees that grew up along a fence row and were left when the fence was removed (fig. 6.15).

Fig. 6.15 "FENCE SHADOW." A double line of old trees along a Woodford County lane remains as a "fence shadow." Rock fences once stood on both sides of the road. Trees grew up along them, and, when the fence on the left side was removed and replaced by a plank fence, the trees remained. Their alignment indicates where the fence once stood.

Rock Fence Preservation

As cultures and technologies change, or as one culture group succeeds another, people imprint the land with survey lines and boundaries, routeways, and buildings. They etch their culture, generation by generation, into the face of the land. Often their only record is what they have left on the landscape, a chronicle of how they lived and worked. Deciphering this imprint shows that current residents are merely the most recent habitants of a land that has hosted a sequence of occupants, all leaving their marks.

People deal with the remnants from these past eras in a variety of ways. They remove some remnants when they get in their way or their appearance displeases. Some common artifacts, such as houses, crossroads markets, or farm buildings, seem to possess little historic merit and are removed to increase convenience or to make way for new structures that serve a different technology. Some landscape features—old churches and graveyards, for example—resist change and may stand for generations. People ignore others because they are unobtrusive or not in the way of change, and, like the small town bypassed by a rerouted highway, they continue to serve a local constituency while the new age speeds past.

The rock fences of the Bluegrass have faced all of these circumstances. Changes in road-building technology have eliminated some, while farming priorities have taken others. Some remain in out-of-the-way places, ignored and in disrepair. Still others border favored old roads, are venerated, and become the stuff of legend. Myths arise to explain their presence and meaning.

Our society does aspire to preserve some of the artifacts of the past, but we do this selectively. It is simpler to argue that a great building, designed by a great architect or occupied by a great man, merits preservation than it is to argue that a neighborhood market or a row of common working people's houses are worthy of the same effort. The unfortunate result of such selective preservation is that the geographic, historic, and cultural past is lost, and what is left is a unique artifact out of context amidst other structures to which it has no functional attachment. Such artifacts tell very little of their place in events during the time they flourished.

Selective preservation, further, conveys a very biased sense of the past. Such preservation is undemocratic, focusing upon only a small segment of our heritage—the rich and powerful—rather than its totality. We thus lose both the sense of attachment to place and region that the builders experienced and the cultural context within which they lived and worked. If large areas of the landscape are not kept intact so that the functional and spatial relationships among the landscape's elements are apparent, little is preserved, and the visual chronicle is lost.

In addition to the story that rock fences tell of the process of migration and the diffusion of ideas and artifacts, the fences have rich aesthetic qualities. For two centuries, the Bluegrass countryside has been coveted land. Today it contains the state's largest urban and industrial centers—Louisville, Lexington, and Covington-Newport. Land valued for agriculture and for scenery has been taken over by urban uses, yet much of it remains in farms and retains the pastoral qualities that rural people and city dwellers enjoy. Its towns lie, as in no other part of the country, within a great park of creeks and grassy woodlands. Rock fences that have been in place a century or more still line some country roads and lanes. The beauty of gray rock fences crossing green pastures appeals to our aesthetic sense and contributes to appreciation of the geographic and historic context that gives this region distinction among American places. Rock fences have become a poignant reminder of the skilled craftsmanship of Scottish, Irish, and black masons, and of farmers whose labor and ideas created the human landscape of the Bluegrass.

The Value of Countryside

This work has made much of the geographic connections between Kentucky's Bluegrass landscape and that of the United Kingdom and Ireland, and it seems fitting to make a final link here. Two passages describe the context of British pastureland and rock walls and their simple beauty. A highly regarded tour guide wrote: "Field patterns and the networks of walls and hedgerows which trace out the mosaic are the essence of rural beauty. There is no single facet of the British landscape which is more valuable than this intricate tracery in grey and green" (Muir and Duffey 1984, 31).

And an author known for his books and television series on the life of a country veterinarian in Yorkshire wrote: "The Dales haven't changed. The dry stone walls still climb up the hillsides as they have always done. Those wonderful walls, often the only sign of the hand of man, symbolize the very soul of the high Pennines, the endlessly varying pattern of grey against green, carving out ragged squares and oblongs" (Herriot and Brabbs 1979, 22-24). James Herriot, Richard Muir, and Eric Duffey are accomplished observers and commentators on interpreting landscapes. They appreciate the land as both the context for life and as a visual tableau of past and present structures set upon a varied topography.

The contemporary British rural landscape is the product of the efforts of hundreds of generations. Inhabitants of Great Britain revere its beauty and salutary effect. Its visual qualities are portrayed through the words of David Lowenthal and Hugh Prince:

> The English landscape is altogether so tamed, trimmed, and humanized as to give the impression of a vast ornamental farm, as if the whole of it had been designed for visual pleasure. Hedgerows, stonewalls, and roads contain vistas, model contours, reinforce contrasts between textures and colors of adjacent fields, and link contrasting landscapes, as in the Yorkshire Dales, where limestone walls, "unbroken and continuous from every tram terminus to the last wilderness of bog and cloud," carry the eye from grassy valley bottom up to untenanted heath and rocky cliff. Riversides and roadsides are trimmed and grass-verged. . . . Although it has few strong vertical lines, the English landscape looks both architectural and tidy. [1964, 325]

In stark contrast, Lowenthal found the American landscape occupied by farms and villages that were "collections of heterogeneous buildings marooned in wastelands" with little concern for context and where "the art of making a pattern in the environment is entirely neglected" (1968, 82).

The British recognize that there is utility in the aesthetics of their rural landscape. J.A. Patmore points out, "Few tracts of rural America have the intimately meandering and dense network of rural roads which characterizes so much of Britain, with the gentle backcloth of an immensely varied, mature agricultural landscape" (1974, 82).

In central Kentucky, such landscapes need not be invented; they have been in place for a century or more. This rural zone surrounding the city can be an integral part of the urban enviroment, rich in potential to meliorate the life there. The irony is that the Bluegrass landscape, though a small portion of the American whole, has long been favorably compared with England, and many of its landscape features were built by people who were either of British Isles heritage or loved things British. Its traditions, artifacts, and unique physical venue set it apart from the rest of the country, circumstances well understood by its important tourist industry but often unprotected from the forces of "economic development."

Kentuckians appreciate the Bluegrass as a great woodland park. On a countryside made famous by its thoroughbreds and other blooded horses, the rock fences along the roadways are one of the most distinguished features and are enjoyed by viewers who are not knowledgeable in their history as well as by those who are.

Many people recognize the rock fences as a valuable asset. A letter to the editor of the *Lexington Herald-Leader,* entitled by the editor "Law is needed to protect rock fences," expressed a common feeling. It read, "This letter is in defense of our beautiful rock fences, quickly being destroyed like many of the things that have made our state unique. I've seen so many of these fences removed just to widen roads. . . . I'd like to see a law passed . . . to make every effort to keep the few remaining fences. When they are finally gone, they will be gone forever" (Placilla 1988).

Legal precedents for preserving rock fences can be found. For example, the zoning board at Pawling, New York, enforces regulations that prohibit destruction of stone walls in an area called Quaker Hill (Straight 1987, 70). In Kentucky, public agencies are studying protection methods. The Kentucky Heritage Council has completed a how-to plan for evaluating rock fences and nominating them to the National Register of Historic Places. The Lexington-Fayette Urban County Government planning staff received a grant from the Kentucky Division of Forestry to map the rock fences along the roadways of Fayette County as a step toward their conservation and has recommended that Fayette County rock fences be declared historic landmarks. The mayor of Lexington has appointed a greenspace committee to draft an enabling ordinance for a greenspace preservation and restoration commission, with preservation of rock fences as one of its broad view objectives.

Landscape preservation and maintenance are no simple tasks, either in cost or in method. People in the United Kingdom have worked at preservation and conservation and can provide a model for us to follow. In Scotland and northern England, rock fence masons and their craft were nearly extinct by the 1950s. Then Frederick Rainsford-Hannay of southwest Scotland wrote his book on dry stone walls and began lecturing on the craft of proper wall building. Rainsford-Hannay's preservation group, now the Dry Stone Walling Association, has been joined by other groups that have a committed interest in maintaining the walls and dykes of Great Britain. These organizations and numerous local associations involved with rural preservation were successful in reviving dry stone walling as a skilled occupation (Dry Stone Walling Association 1989; Shaughnessy, 1986, 38).

Hundreds of people each year get an opportunity to acquire stonewalling skills and to use those skills repairing the many fences that still stand in Scotland, Yorkshire, and the Lake District (Garner 1984, 29). The cost of building new walls may run to ten or fifteen pounds (or $15.00 to $27.50 in 1990 dollars) per running yard, prices too high for most wall owners. Government grants of up to 50 percent of the cost of the work make stone about the same price as other forms of fencing. Masons trained by the Dry Stone Walling Association now rebuild about thirty miles of dry stone walling each year in southwest Scotland, funded, in part, by grants (Tufnell 1989). In the British Isles today, stone walls are regarded as national treasures.

Relatively few of Kentucky's rock fences remain. Repairing them should be a minor industry in the Bluegrass; yet, for reasons that are easy to understand, it is not. Most landowners do not have the time to learn how to rebuild their fences properly. They may be stymied by the cost of hiring masons who know and use the appropriate methods, and they may not know where to find a competent mason. Aside from the fence building demonstrations in New England conducted by the Dry Stone Walling Association of Great Britain, we know of no effort in America aimed at preserving the stonemasons' craft or the fences that the British, the Irish, and the freedmen built.

How people perceive the value of rock fences depends upon their perspective. The farmer who must spend weeks in the hot sun repairing his fences may question their practicality. The landowner who must hire a mason to build new fences or rebuild fallen sections may think the costs too high. The city dweller may want the fences preserved because they enhance the beauty and character of the Bluegrass countryside. We hope that the temporal and spatial contexts for the region's rock fences provided in this book will promote efforts to preserve those that still survive.

Boyle County

Mercer County

Scott County

Clark County

Mercer County

Scott County

Anderson County

Scott County

Fayette County

Boyle County

Mercer County

Woodford County

Woodford County

Woodford County

APPENDIX 1

Historic and Contemporary Fence Masons of Central Kentucky

Few archival resources are as valuable to historical research as those containing dates and places when and where people lived and worked. Such records are essential in tracing the movement of people and ideas from the Old World to the New and from one American place to another. This appendix lists in alphabetical order the names of men of all ethnic groups who are responsible for Kentucky's rock fences and is a reference to the cultural origins of these nineteenth- and twentieth-century craftsmen.

These names of fence masons and quarrymen derive from manuscript collections, books, periodicals, correspondence, personal interviews, and U.S. Census records. Every person in this appendix was a "stonemason," "fence mason," or "rock fencer" unless otherwise indicated. The locations in parentheses after some names are birthplaces. In some censuses the birthplaces of the respondents' parents were recorded, in which case that information is in parentheses in addition to the masons' birthplaces. If the birth or death date is not known, the date the person worked is indicated. The county listed after the dates is the county in which the person resided or, if that is not known, the county in which he worked. The source of the information appears in parentheses, with details about the source in the Reference List. For some names, additional information that helps explain relationships and associations is bracketed. If the mason was not the head of the household, the person who was the head of the household is noted; this implies that the mason boarded with, was related to, or worked for that person.

There are cautions about the accuracy of this list: names are spelled as in the censuses, and these are often phonetic spellings; therefore, the same person may be listed under a different spelling in a different year. Also, some of the stonemasons whom the census takers noted by occupation may have been retired at the date of the census. Simeon Smith, for example, was ninety years old and blind at the time of the 1850 census, yet his occupation is given as stonemason. Dates of birth are very imprecise. If the census taker interviewed only one person in a household, the respondent may have given an approximate date of birth for boarders or employees; this is why a person's birth date may vary from one census to the next. Some census schedules are difficult to read, making accurate spellings of names impossible; and some names may have been overlooked.

The Brutus J. Clay records demonstrate that a minority of stonemasons dwelt in one place long enough to be included in census counts. Of the twenty-four stonemasons who worked for Clay through the years, only thirteen were ever listed on any census. If this circumstance is typical, the Bourbon County stonemasons listed in this appendix represent about one-third of the total number of stonemasons working in Bourbon County between 1850 and 1910. This list therefore may represent approximately one-third of the total number of Bluegrass fence masons for the counties and years the censuses were read, and only a small fraction of those who worked in other Kentucky counties.

If the census taker listed a person as a "stone cutter," we did not include him in this list of stonemasons. Stone cutters, stone sawyers, stone setters, marble cutters, and marble workers often lived in the household or neighborhood of a marble agent, tombstone shopkeeper, or manufacturer of monuments. Stone cutters, therefore, were probably tombstone carvers, not fence builders. The census taker recorded some people simply as "fencer" or "fence builder"; these also are not included in the list of names of stonemasons since they could have been post-and-rail or wire fence builders. It is very likely that some people listed as stonemasons built not fences but stone houses or stone foundations for wooden and brick houses. This distinction cannot be made from most census records, so all people listed as stonemasons are included in this list, unless the 1910 census specified that they were working on a structure other than a rock fence (the respondent's workplace, in addition to his occupation, was first recorded in the 1910 census). The 1890 census was destroyed by fire.

There are additional cautions about the contents of this list: slaves were not enumerated by occupation, and all blacks were first listed in the 1870 census. Since we read the censuses for only five counties—Anderson, Bourbon, Mercer, Scott, and Woodford—through 1910, the largest number of masons recorded here are from those five counties; the censuses of other Kentucky counties after 1850 contain additional masons' names not in this appendix. Many stonemasons were itinerant craftsmen and were not counted by census takers. The list also greatly underrepresents contemporary fence masons; we did not systematically seek the names of all working masons but simply listed those we encountered, observed, or interviewed. Abbreviations used in the list are as follows:

b = born
CMW = 1975 file of author
Co. = county
d = died
fa = father
fam = family
IRE = Ireland
mo = mother
occ = occupation
par = parents
USC = U.S. Census (date and county can be determined by entry)
w = date working

Blacks

Adams, Will (KY): b 1822 w 1870 Scott Co. (USC)
Adams, William (KY): b 1820 w 1880 Scott Co. (USC)
Alexander, Elisha (KY): b 1853 w 1900, 1910 Scott Co. (USC)
Alexander, Wm. (KY): b 1878 w 1900 Scott Co. (USC) [son of Elisha Alexander, above]
Anderson, Harry: b 1880 w 1900 Woodford Co. (USC) [listed with Lewis Anderson, below]
Anderson, Lewis: b 1862 w 1900 Woodford Co. (USC) [in 1910 listed as a stonemason on buildings]
Anderson, Wesley: b 1871 w 1910 Bourbon Co. (USC)
Andumn, Louis: w last part 1800s Woodford Co. (Thierman 1959)
Asher, James: b 1858 w 1900, 1910 Bourbon Co. (USC)
Bargo or Bango, John (KY): b 1840 w 1880 Scott Co. (USC)
Beovse?, George (KY): b 1809 w 1870 Scott Co. (USC)
Black, Amos: w 1941 Woodford Co. (Landau 1941; Guy 1990) [taught by John Guy, below; kin to Guy family]
Black, Richard (KY): b 1860 w 1900 Mercer Co. (USC)
Bownal, T. (Ky): b 1860 w 1900 Mercer Co. (USC)
Branham, Antony (KY): b 1827 w 1880 Woodford Co. (USC)
Brannun?, Anthony (KY): b 1818 w 1870 Woodford Co. (USC)
Brooks, Isaac (KY): b 1802 w 1870 Scott Co. (USC)
Brown, Benjamin H.: b 1892 w 1910 Bourbon Co. (USC) [listed with Henry and Ralph Brown, below]
Brown, Henry: b 1863/1875 w 1900, 1910 Bourbon Co. (USC) [lived next door to Ralph Brown, below, in 1900; listed with Benjamin H. and Ralph Brown, above and below, in 1910]
Brown, Howard: b 1827 w 1880 Bourbon Co. (USC)
Brown, Howard: b 1857 w 1900 Bourbon Co. (USC)
Brown, Jim: Franklin Co. (Guy 1990)
Brown, Ralph: b 1857 w 1900, 1910 Bourbon Co. (USC) [listed with Henry and Benjamin H. Brown, above, in 1910]
Buckner, Horace: b 1818/1830 w 1870, 1880 Bourbon Co. (USC) [listed 1870 with brother-in-law Domenick Warren, day laborer]
Buford, Jeremiah: b 1803 w 1850 Fayette Co. (USC)
Bunnell? Bunnett?, Anthony: b 1795 w 1850 Fayette Co. (USC)
Burly, Haey or Hacy (KY): b 1829 w 1870 Scott Co. (USC)
Bush, Daniel: b 1832 w 1880 Bourbon Co. (USC)
Campbell, Vinse: b 1823 w 1880 Woodford Co. (USC)
Carlyle, Henry: b 1830 w 1900 Bourbon Co. (USC)
Carr, George: b 1845 w 1880 Woodford Co. (USC)
Carron, John: w 1900s Boyle Co. (Caldwell 1989)
Carter, ("Uncle") John: w 1940s Woodford Co. (Gormley 1987) [worked with James Rogers Gormley, below]
Chenault, John: w early 1900s Bourbon Co. (Thierman 1959)
Chipman, Adam (VA): b 1789 w 1850 Scott Co. (USC)
Clarkr?, Jno (KY): b 1859 w 1900 Scott Co. (USC)
Clay, D.C.: w 1873 Bourbon Co. (Clay 1846-1877)

Coffee, Thomas: b 1874 w 1900 Bourbon Co. (USC)
Collins, S.: b 1850? w 1880 Bourbon Co. (USC)
Colman, Wm. (KY): b 1825 w 1870 Scott Co. (USC)
Conner, Edward: w 1989 Bourbon Co. (Hinkle 1988)
Cotton, John: b 1805 w 1850 Fayette Co. (USC)
Crittenden, J. (KY): b 1816 w 1870 Scott Co. (USC)
Cunningham, Charles (KY): b 1844 w 1870 Mercer Co. (USC)
Custar, Green: b 1830 w 1870 Bourbon Co. (USC)
Dedman, George: w 1970s Fayette Co. (Lee 1984; Guy 1990) [kin to the Guy family]
Depp, James (KY): b unknown w 1900 Mercer Co. (USC)
Dickerson, A.: b 1814 w 1850 Shelby Co. (USC)
Divine, Milton: b 1820 w 1880 Woodford Co. (USC)
Dobson, Mort B.: b 1866 w 1900 (quarry laborer) Woodford Co. (USC) [listed with Elija Mason and William Smith, below]
Dodd, George: w 1910 Woodford Co. (USC)
Dudley, Moreell (KY): b 1865 w 1900 Scott Co. (USC)
Dudley, Toby (KY): b 1815 w 1870 Scott Co. (USC)
Duncan, James B.: b 1868 w 1910 Bourbon Co. (USC)
Duncan, Jesse: b 1886 w 1910 Bourbon Co. (USC) [son of James B. Duncan, above]
Duncan, Josh (KY): b 1831 w 1880 Mercer Co. (USC)
Easton, William (KY): b 1839 w 1860 Mercer Co. (USC)
Edwards, Ben: b 1841 w 1900 Woodford Co. (USC)
Ewing, Charles: b 1825 w 1860 Bourbon Co. (USC) [listed with Peter Hedges, white, farmer]
Fenix, George (KY): b 1868 w 1910 Scott Co. (USC) [probably the same person as George Finex, below]
Field, G____: b 1835 w 1880 Woodford Co. (USC)
Finch, Charles: b 1848 w 1880 Bourbon Co. (USC)
Finch, W. Henry: b 1864 w 1900, 1910 Bourbon Co. (USC)
Finex, George: b 1865 w 1900 Woodford Co. (USC) [listed with Elija Watson, Samuel Guy, and Berry Scruggs, below]
Fish, Joe: w 1940s Bourbon Co. (J. Soper 1988)
Fowler, Thornton (KY): b 1818/1820 w 1870, 1880 Scott Co. (USC)
French, Charles: b 1845 w 1870 Bourbon Co. (USC)
French, George: b 1846 w 1910 Bourbon Co. (USC)
Gaines, Peter (KY): b 1835 w 1900 (quarrier) Scott Co. (USC)
"Garland": b 1808 w 1850 Bourbon Co. (USC)
Garland, Winston: b 1810 w 1870 Bourbon Co. (USC) [may be same person as "Garland"]
Gaskine, John (KY): b 1844 w 1870 Scott Co. (USC)
Gaskins, John (KY): b 1834 w 1900 Scott Co. (USC)
Gaskins, John (KY): b 1843 w 1880 Scott Co. (USC)
Gay, George: b 1850 w 1880 Bourbon Co. (USC)
Gentry, Pete: w early 1900s Bourbon Co. (Thierman 1959)
Gill, George: b 1876 w 1900 Woodford Co. (USC)
Gill, James: b 1852 w 1880 Woodford Co. (USC) [brother of John Gill, below]
Gill, Jim: w 1940s Woodford Co. (Gormley 1987) [worked with James Rogers Gormley, below]
Gill, John: b 1854 w 1880 Woodford Co. (USC) [brother of James Gill, above]

Green, Andrew J.: b 1847 w 1900 Bourbon Co. (USC)
Green, J.: b 1841 w 1880 Bourbon Co. (USC)
Green, J.: b 1847 w 1880 Bourbon Co. (USC)
Green, Richard: b 1825 w 1850 Boyle Co. (USC)
Green, Sandy: b 1865 w 1900 Woodford Co. (USC)
Green, Stanley: b 1886 w 1910 Bourbon Co. (USC)
Green, Thomas: b 1820 w 1850 Boyle Co. (USC)
Green, William: b 1874 w 1910 Bourbon Co. (USC)
Griffin, Gabe (KY): b 1839 w 1870 Scott Co. (USC)
Gunn, M.: b 1820 w 1850 Boyle Co. (USC)
Guy, Ben: w early 1900s Woodford Co. (Thierman 1959) [son of Henry Guy, below]
Guy, Charlie: Franklin Co. (Guy 1990) [son of Ben Guy, above; grandson of Henry Guy, below]
Guy, Clarence: w 1980s Fayette Co. (Lee 1984)
Guy, Clarence, Jr.: w 1989 Fayette Co. (Vanderhoef 1989)
Guy, Clay: w early 1900s Woodford Co. (Thierman 1959) [son of Henry Guy, below]
Guy, Ed: b 1885 d 1959 Woodford Co. (Thierman 1959) [son of Robert Guy and grandson of Henry Guy, below]
Guy, Edward: w 1941 Shelby Co. (Landau 1941) [nephew of John Guy, below]
Guy, Frank Jr.: w mid-1900s Bourbon Co. (Thierman 1959) [son of Louis Guy, below?]
Guy, Frank E., Sr.: b 1905 d 1987 Fayette Co. (*Lexington Herald-Leader* 17 July 1987) [son of Ed Guy, above, and great-grandson of Henry Guy, below]
Guy, Henry: w as a slave Woodford Co. (Thierman 1959)
Guy, James B: Franklin Co. (Guy 1990) [son of Ben Guy, grandson of Henry Guy, above]
Guy, Jim B.: Franklin Co. (Guy 1990) [son of John Guy Sr., below]
Guy, John, Jr.: w 1941 Franklin Co. (Landau 1941) [son of John Guy, below]
Guy, John, Sr.: w 1941, 1952 Franklin Co. (Landau 1941; O'Conner 1952)
Guy, John Henry, III: w 1989 Franklin Co. (Guy 1989) [son of John Guy Jr., above; grandson of John Guy, above]
Guy, Lewis: b 1866 w 1900 Woodford Co. (USC) [probably the same as Louis Guy or Lewis Guy, below]
Guy, Lewis: Franklin Co. (Guy 1990) [son of John Guy, Sr.]
Guy, Louis: w 1906-1956 Fayette Co. (R.R. 1956; Thierman 1959) [son of Henry Guy, above]
Guy, Nathan: w 1967 Fayette Co. (Williamson 1967)
Guy, Nathaniel: w first ½ 1900s Bourbon Co. (Guy 1989) [son of Louis Guy and great-uncle of John H. Guy III, above]
Guy, Nathaniel: Franklin Co. (Guy 1990) [son of John Guy, Sr., above)
Guy, Robert: w early 1900s Woodford Co. (Thierman 1959) [son of Henry Guy, above]
Guy, Robert: b 1839 w 1880 Woodford Co. (USC) [probably the same as Robert Guy, above]
Guy, Samuel: b 1872 w 1900 Woodford Co. (USC)
Guy, Samuel: b 1873 w 1900 Woodford Co. (USC) [listed with Eugene Watson, Berry Scruggs, George Finex, above and below; may be same person as Samuel Guy, above, counted twice in the same census]
Guy, Shedrick: w early 1900s Woodford Co. (Thierman 1959) [son of Henry Guy, above]
Haffort, Rob (KY): b 1835 w 1870 Scott Co. (USC)

Hale? Hall?, Thomas: b 1808 w 1850 Boyle Co. (USC) [listed with Francis Sarcens? Sarcene?, white]
Hamilton, George: b 1841 w 1900, 1910 (quarrier) Bourbon Co. (USC)
Hardue, Priestly (KY): b 1853 w 1870 Mercer Co. (USC)
Harlan, Henry (KY): b 1820 w 1870 Mercer Co. (USC)
Harris, John: Franklin Co. (Guy 1990) [kin to the Guy family]
Hayden, Mexico: w 1989 Jessamine Co. (Tate 1984)
Henderson, George: b 1883 w 1910 Bourbon Co. (USC)
Hernix [Fenix?], Wyat (KY): b 1830 w 1880 Scott Co. (USC)
Hicks, Robert: w 1970s Clark Co. (Niles 1984; Venable 1989)
Higgins, Burl (KY): b 1854 w 1900 Scott Co. (USC)
Higgins, Henry: near retiring 1989 Fayette Co. (Guy 1989) [kinsman of Guy brothers, above]
Higgins, Jack: near retiring 1989 Fayette Co. (Guy 1989) [kinsman of Guy brothers, above]
Higgins, Stanley: w 1980s Fayette Co. (Niles 1984)
Hightower, Milton (KY): b 1830 w 1870 Mercer Co. (USC)
Hogan, Robert, Sr.: retired 1989 Franklin Co. (Guy 1989)
Hosen?, Sam (KY): b 1820 w 1880 ("work in rock") Scott Co. (USC)
Hoskens, Alonzo: w 1980s Fayette Co. (Casey 1988)
Howard, John (VA): b 1800 w 1860 Scott Co. (USC)
Huffman, Mason: b 1866 w 1910 Bourbon Co. (USC)
Hutzle, Andrew: b 1873 w 1900 Bourbon Co. (USC)
Hutzle, David M.: b 1847 w 1910 Bourbon Co. (USC)
Hutzle, Elmer: b 1881 w 1900 Bourbon Co. (USC) [brother of Andrew Hutzle, above]
Hutzle, W. Edward: b 1875 w 1910 Bourbon Co. (USC)
"Jabo": w 1940s Woodford Co. (Gormley 1988)
Jackson, George: b 1854 w 1900 Woodford Co. (USC)
Jackson, John: b 1845 w 1910 (quarry laborer) Woodford Co. (USC)
Jackson, John Bill: w 1940s Woodford Co. (Gormley 1988) [worked with James Rogers Gormley, below]
Jacobs, William C., Jr.: w 1970s, 1980s Franklin Co. (Guy 1990; Lee 1984)
Jacobs, William C., Sr.: b c1900 w 1970s, 1980s Franklin Co. (Guy 1990; Lee 1984)
Jamison, Matthew: retired before 1989 Franklin Co. (Guy 1989)
Jockey, Pete: w 1940s-1960s Bourbon Co. (T. Soper 1988)
"Joe": w 1867 Bourbon Co. (Clay 1854-1875)
Johnson, Caesar: b 1830 w 1860 Mercer Co. (USC)
Johnson, Clay J.: b 1883 w 1910 Bourbon Co. (USC)
Johnson, Reuben: b 1839 w 1880 (quarrier) Woodford Co. (USC)
Johnson, Thomas: b 1851 w 1910 Bourbon Co. (USC)
Johnson, Vernon: w 1989 Franklin Co. (Guy 1989) [taught by John H. Guy III, above]
Johnson, Walker: b 1850 w 1880 Bourbon Co. (USC)
Johnson, Willis: b 1820 w 1850 Jessamine Co. (USC)
Jones, Elijah (KY): b 1869 w 1900 Scott Co. (USC)
Jones, George: b 1883 w 1900 Bourbon Co. (USC) [son of Pleasant Jones, below]
Jones, Pleasant: b 1856 w 1900 Bourbon Co. (USC) [in 1910 was a stonemason on houses, so probably not a fence mason; son George Jones, above, probably not either]
Kelley, Thomas: b 1840 w 1870 Bourbon Co. (USC)

Kenney, Pete: b 1840 w 1880 Bourbon Co. (USC)
Lee, John: w late 1800s Woodford Co. (Cause 1980)
Lee, Legrand: b 1893 Woodford Co. (Cause 1989) [son of John Lee, above]
Lee, Richard: b 1810 w 1870 Bourbon Co. (USC)
Lee, Robert: b 1829 w 1870 Woodford Co. (USC)
Lee, Stephen: b 1908 d 1987 Woodford Co. (Cause 1980) [son of John Lee, above]
Lowden, John: w 1940s Bourbon Co. (J. Soper 1988)
Lowry, John: b 1878 w 1900 (quarrier) Woodford Co. (USC)
Marshall, Crawford: b 1839 w 1870 Bourbon Co. (USC)
Marshall, John: b 1889 w 1910 Woodford Co. (USC)
Martin, Sam (KY): b 1874 w 1900 (quarrier) Scott Co. (USC)
Martin, William: b 1867 w 1910 Bourbon Co. (USC)
Mason, Elija: b 1867 w 1900 (quarry laborer) Woodford Co. (USC) [listed with William Smith and Mort B. Dobson, above and below]
McClellan, William, Jr.: Franklin Co. (Guy 1990) [son of William McClellan, Sr., below; kin to the Guy family, above]
McClellan, William, Sr.: Franklin Co. (Guy 1990) [kin to the Guy family, above]
McCray, Mack: b 1855 w 1910 Bourbon Co. (USC)
McGown, Willis: b 1836 w 1900 Bourbon Co. (USC)
Meaux: w late 1800s Boyle Co. (Caldwell 1989)
Meaux, Robin/Rubin: b 1796/1802 w 1860, 1870 Mercer Co. (USC)
"Melborn": w 1848 Bourbon Co. (Clay and Thornton 1848)
Melvin, Richard (VA): b 1790 w 1860 Mercer Co. (USC)
Miller, Charles: w 1990 Fayette Co. (Rebmann 1988)
Miller, William: w 1970s Fayette Co. (Guy 1990; Lee 1984)
Mim?, Jack (KY): b 1812 w 1880 Mercer Co. (USC)
Minor, Harry: b 1878 w 1900 (quarrier) Woodford Co. (USC)
Morgan, George (KY): b 1852 w 1900 Mercer Co. (USC)
Myers, John: b 1840 w 1880 Woodford Co. (USC)
Nelson, Henry: b 1849 w 1910 Bourbon Co. (USC)
Offutt, Chris: w late 1800s Scott Co. (Bevins 1988)
Oliver, Washington (KY): b 1820 w 1870 Scott Co. (USC)
Parish, Elias: b 1852 w 1900 (quarry worker) Woodford Co. (USC)
Patterson, Robert: b 1878 w 1900 (quarrier) Woodford Co. (USC)
Payne, Robert (KY): b 1812 w 1870 Mercer Co. (USC)
Peacock, Bill: w 1980s Fayette Co. (Tate 1984)
Phenix, Nich (KY): b 1832 w 1870 Scott Co. (USC) [probably same surname as Finex, above, and Phoenix, below]
Phillips, James (KY): b 1858 w 1900 (quarrier) Scott Co. (USC)
Phoenix, George (KY): b 1860 w 1900 Scott Co. (USC)
Pierce, Will: b 1858 w 1900 Bourbon Co. (USC)
Prentiss/Prentice bros.: w 1980s Jessamine Co. (Tate 1984)
Raglin, Charles: b 1887 w 1910 Woodford Co. (USC)
Redd, George: b 1838 w 1880 Woodford Co. (USC)
Reed, Joseph: b 1786 w 1850 Boyle Co. (USC)
Richardson, John: b 1841 w 1880 Bourbon Co. (USC)
Richardson, Thomas: b 1837 w 1880 Bourbon Co. (USC)
Roberson, Henry: b 1830 w 1870, 1880 Bourbon Co. (USC) [listed with Peter Lewis, black, farm laborer]

Robinson, Gerry (KY): b 1850 w 1900 Scott Co. (USC)
Robinson, Wallace: b 1901 Fayette Co. (Mead 1978; Tate 1984)
Roland, Neal (KY): b 1851 w 1900 Scott Co. (USC)
Rozor, John: b 1862 w 1910 Bourbon Co. (USC)
Samuals, Gray, Jr. (KY): b 1881 w 1900 Scott Co. (USC) [son of Graym Samuals, below]
Samuals, Graym (KY): b 1832 w 1900 Scott Co. (USC)
Samuals, Harvey (KY): b 1877 w 1900 Scott Co. (USC) [son of Graym Samuals, above]
Samuals, Prewill (KY): b 1878 w 1900 Scott Co. (USC) [son of Graym Samuals, above]
Samuals, Wm. J. (KY): b 1871 w 1900 Scott Co. (USC) [son of Graym Samuals, above]
Samuels, Gaanule? (KY): b 1830 w 1870 Scott Co. (USC)
Samuels, Henry (KY): b 1862 w 1880 Scott Co. (USC)
Samuels, J. (KY, fa KY): b 1861 w 1900 Scott Co. (USC)
Samuels, Thomas (KY): b 1868 w 1900 Scott Co. (USC)
Saunders, Afton (KY): b 1841 w 1870 Mercer Co. (USC)
Scruggs, Berry: b 1857 w 1900 Woodford Co. (USC) [listed with Eugene Watson, Samuel Guy, George Finex, above and below]
Scruggs, Issaac: b 1826 w 1880 Woodford Co. (USC)
Scrugs?, Peter: b 1831 w 1880 Woodford Co. (USC)
Seawright: w early 1900s Boyle Co. (Caldwell 1989)
Senugga?, Barry (KY): b 1858 w 1900 Scott Co. (USC)
Sinclair, John (KY): b 1837 w 1900 Scott Co. (USC)
Smith, George: w 1915-1920 Scott Co. (Hall 1989)
Smith, Spiden (MS, fa KY): b 1857 w 1900 (quarrier) Scott Co. (USC)
Smith, William: b 1863 w 1900 Woodford Co. (USC) [listed with Elija Mason and Mort B. Dobson, quarry laborers, above]
Spears, Bill: w mid 1900s Harrison and Scott Co. (Bevins 1988)
Stamps, Andrew G.: b 1845 w 1910 Bourbon Co. (USC)
Stevenson, ?: b 1805 w 1880 Bourbon Co. (USC)
Stevenson, Caesar: w 1910 Bourbon Co. (USC)
Stowers, Samuel: b 1790 w 1850 Harrison Co. (USC)
Stewart, Tom: b 1864 w 1900 Bourbon Co. (USC)
Tarry, William (KY): b 1842 w 1880 Scott Co. (USC)
Taylor, Ed: Fayette Co. (Guy 1990) [first cousin to John Henry Guy III]
Taylor, George (KY): b 1846 w 1900 Scott Co. (USC)
Taylor, Ham: b 1835 w 1900 Bourbon Co. (USC)
Taylor, Homer Ed: w 1988 Woodford Co. (Guy 1989) [cousin of John H. Guy III, above]
Thomas, Buford: b 1843 w 1870 Woodford Co. (USC)
Tolbert, Robert (KY): b 1830 w 1880 Scott Co. (USC)
Toliver, Haze: b 1829 w 1880 Woodford Co. (USC)
Trotter, Wilson: w 1900 Bourbon Co. (USC)
Trumbo: w late 1800s Boyle Co. (Caldwell 1989)
Trumbo, Bill: w 1970s Fayette Co. (Guy 1990; Lee 1984)
Turner, Samuel: b 1949 w 1900 Bourbon Co. (USC)
Uttley, Add (KY): b 1852 w 1900 Mercer Co. (USC)
Vincent?, Garvin?: b 1805 w 1860 Bourbon Co. (USC)
Walker, Daniel? (KY): b 1817 w 1870 Mercer Co. (USC)
Walker, George (KY): b 1851 w 1900 Scott Co. (USC)
Walker, Thomas (KY): b 1815 w 1870 Mercer Co. (USC)
Ware, Alfred: b 1865 w 1910 Woodford Co. (USC)

Ware, William: b 1896 w 1910 Woodford Co. (USC) [son of Alfred Ware, above]
Washington, George: b 1827 w 1870 Bourbon Co. (USC)
Washington, Pompii: b 1840 w 1880 Anderson Co. (USC)
Watson, Eugene: b 1872 w 1900 Woodford Co. (USC) [listed with Samuel Guy, Berry Scruggs, and George Finex, above]
Wesley, George: b 1831 w 1910 Bourbon Co. (USC)
Whaley, Jerry: b 1826 w 1880 Bourbon Co. (USC)
White, John (KY): b 1872 w 1900 Mercer Co. (USC)
White, Thomas (KY): b 1865 w 1900 Mercer Co. (USC)
Williams, Ch. (KY): b 1842 w 1880 Mercer Co. (USC)
Williams, Gren (KY): b 1823 w 1880 Scott Co. (USC)
Williams, John (KY): b 1866 w 1880 Scott Co. (USC) [son of Gren Williams, above]
Williams, Noah (KY): b 1861 w 1880 Scott Co. (USC) [son of Gren Williams, above]
Williams, Sam (KY): b 1863 w 1880 Scott Co. (USC) [son of Gren Williams, above]
Williams, William: b 1833 w 1870 Bourbon Co. (USC)
Williams, Willis: b 1830 w 1880 Woodford Co. (USC)
"Woodson": w 1851 Bourbon Co. (Clay 1839-1853)
Woodson, Henry: b 1850 w 1880 Woodford Co. (USC)
Yates, Sam: b 1842 w 1880 Anderson Co. (USC)

Whites

Adams, Thomas (KY): b 1820 w 1850 Jessamine Co. (USC)
Aendle?, Michael (IRE): b 1810 w 1860 Scott Co. (USC)
Alcorn, Will (KY): b 1821 w 1850 Garrard Co. (USC)
Algood, Isaac G. (NC): b 1827 w 1850 Boyle Co. (USC)
Allen, David (KY): b 1816 w 1870 Mercer Co. (USC)
Allen, Lewis, (KY): b 1807 w 1850 Woodford Co. (USC)
Allen, Stephen (Germany): b 1820 w 1850 Franklin Co. (USC)
Allison, Ted: w 1970s Nicholas Co. (CMW)
Anderson, John (IRE): b 1827 Scott Co. (Bevins 1984a)
Angleton, William (VA): b 1776 (sic) w 1850 Boone Co. (USC)
Arnold, Richard (KY): b 1810 w 1850 Clark Co. (USC) [listed with Peggy Martin, no occ, b KY]
Arnold, Samuel (VA): b 1797 w 1850 Jessamine Co. (USC)
Asher, John Franklin: b 1868 d 1941 Garrard Co. (Gregory 1990)
Atherson, William A. (fa VA): b 1819 w 1880 Woodford Co. (USC)
Bailey, Bazil (KY): b 1805 w 1850 Harrison Co. (USC)
Baley, Ann [Andrew?] (IRE): b 1831 w 1860 (quarrier) Scott Co. (USC) [listed with Charly Linch and family, below, and 18 other Irish quarriers]
Baley, Charley (IRE): b 1828 w 1860 (quarrier) Scott Co. (USC) [lived with Charly Linch and family, below, and 18 other Irish quarriers]
Barber? Barker?, Asah (fa KY): b 1858 w 1880 Anderson Co. (USC)
Barber, Henry: b late 1800s w 1900s Scott Co. (Waugh 1988)
Barker, Leonard (VA): b 1790 w 1850 Harrison Co. (USC)
Barker, Paul: w 1982 Fayette Co. (*Lexington Herald-Leader* 11 May 1982)
Barl____, Jonathan (IRE): b 1824 w 1870 Bourbon Co. (USC)
Barlow, Jacob (KY): b 1819 w 1850 Nicholas Co. (USC)
Barlow, Jesse: w 1839 Bourbon Co. (Clay 1828-1846)

Barmore, Robert: w 1980s Bourbon Co. (L. Waugh 1989)
Barnes, James (VA): b 1784 w 1850 Harrison Co. (USC)
Barrett, Thomas (IRE): b 1814 Scott Co. (Bevins and O'Roarke 1985)
Barton, Dennis (IRE): b 1809 w 1850 Bourbon Co. (USC)
Basil, David (KY): b 1823 w 1860 Mercer Co. (USC)
Bates, John (KY): b 1824 w 1850 Jessamine Co. (USC) [son of Jonathan Bates, below]
Bates, Jonathan (KY): b 1799 w 1850 Jessamine Co. (USC)
Bates, Thomas (KY): b 1828 w 1850 Jessamine Co. (USC) [son of Jonathan Bates, above]
Bates, William (KY): b 1826 w 1850 Jessamine Co. (USC) [son of Jonathan Bates, above]
Baugherty, Thomas (IRE): b 1828 w 1860 (quarrier) Scott Co. (USC) [lives with Charly Linch and family, below, and 18 other Irish quarriers]
Baxter, James (KY): b 1824 w 1850 Jessamine Co. (USC)
Bazzel, Davis (KY): b 1822 w 1880 Mercer Co. (USC) [probably same as David Basil, above]
Beasley, James (GA): b 1798 w 1850 (rock quarrier) Jessamine Co. (USC)
Bell, George (England): b 1830 w 1860 Bourbon Co. (USC)
Bell, Samuel: b 1802 w 1850 Harrison Co. (USC)
Bennett, Willis (KY): b 1803 w 1850 Oldham Co. (USC)
Bennington, William (VA): b 1800 w 1850 Boyle Co. (USC)
Betro, Andrew (Italy): b 1885 w 1910 Woodford Co. (USC)
Birdwistle, Lister (fa KY): b 1875 w 1900 Mercer Co. (USC)
Bishop, John (mo KY): b 1829 w 1850 Mercer Co. (USC) [brother of Thomas H. Bishop, below]
Bishop, Thomas H. (mo KY): b 1825 w 1850 Mercer Co. (USC) [brother of John Bishop, above]
Blackburn, Lewis (KY): b 1810 w 1850 Franklin Co. (USC)
Boden, John (IRE): b 1818 w 1860 Anderson Co. (USC) [listed with Thomas Hanks, farmer, b Ky]
Bolson, Henry (fa Germany): b 1840 w 1900 Bourbon Co. (USC)
Boon?/Bron?, James (KY): b 1798 w 1850 Bourbon Co. (USC) [probably same as James Boone, below]
Boon/Boone, James: w 1837-43 Bourbon Co. (Clay 1828-1846)
Boone, Jonathan: w 1838 Bourbon Co. (Clay 1828-1846)
Bornet, E. (IRE): b 1827 w 1850 Scott Co. (USC) [he and J. McHugh, below, listed with Michael Hickey, laborer, and 9 other laborers, all b IRE]
Boston, William (KY): b 1846 w 1900 Mercer Co. (USC)
Botz, John (Germany): b 1822 w 1850 Mason Co. (USC)
Bowyer, Ezra (KY): b 1814 w 1850 Fayette Co. (USC) [listed w Mrs. Elizabeth Rhodes, b Ky]
Boyle, Hugh (par IRE): b 1864 w 1880 Bourbon Co. (USC) [son of Micheal Boyle, below]
Boyle, Micheal (IRE): b 1825 w 1880 Bourbon Co. (USC)
Bradley, Barney (IRE): b 1822 w 1860 Scott Co. (USC) [listed with Thomas Love, below]
Bradley, Edward (IRE): b 1822 w 1880 Mercer Co. (USC)
Bradley or Bradling, Edward (PA): b 1825 w 1860 Mercer Co. (USC) [listed with James Haggin and Patrick Malary, below]

Bradley, J.R. (KY): b 1845 w 1880 Bourbon Co. (USC)
Bradley, James (IRE): b 1820 w 1860, 1870 Woodford Co. (USC)
Bradley, Paul (IRE): b 1810 w 1860 Mercer Co. (USC)
Bramble? Brumble?, Charles (KY): b 1817 w 1850 Harrison Co. (USC)
Bramble, Elias, Jr. (KY): b 1820 w 1850 Harrison Co. (USC)
Branham, James: b 1963 Scott Co. (B. Waugh 1989) [son of Bobby Waugh; same as James Brannham, below]
Brannham, James: w 1980s Scott Co. (*Lexington Herald-Leader* 20 July 1982)
Brannock, D. (KY): b 1817 w 1850 Shelby Co. (USC)
Brannock, H. (KY): b 1807 w 1850 Shelby Co. (USC) [lives next door to D. Brannock, above]
Brannock, William (KY): b 1801 w 1850 Harrison Co. (USC)
Bratton, Aaron (fa KY): b 1875 w 1900 Bourbon Co. (USC)
Brawner?, B.W. (VA): b 1814 w 1850 Shelby Co. (USC)
Brawner, Robert (MD): b 1789 w 1850 Franklin Co. (USC) [his two sons are brick masons]
Breen, Thomas (IRE): b 1812/1816 d 1892 Scott Co. (Bevins 1984a; Bevins and O'Roarke 1985)
Brian, James (IRE): b 1862 w 1870 Scott Co. (USC)
Briston, John J. (KY): b 1829 w 1850 Franklin Co. (USC)
Brocker, Andrew (Wirtenburg): b 1847 w 1870 Bourbon Co. (USC) [listed with August Gutziet, confectioner, b Prussia]
Brooks, Isaac (KY): b 1808 w 1860 Scott Co. (USC)
Brown, Anderson (KY): b 1822 w 1850 Clark Co. (USC)
Brown, Billy Morgan "Dusty": w 1970s Woodford Co. (Brown 1990) [son of Jesse Brown, Sr., below]
Brown, Henry (KY): b 1818 w 1850 Clark Co. (USC) [lives two houses from Anderson Brown, above]
Brown, James (IRE): b 1823 w 1850 Scott Co. (USC)
Brown, Jeff: w 1980s Woodford Co. (Brown 1990) [son of Billy Morgan Brown, above]
Brown, Jesse, Sr.: b c.1910 d c.1974 Woodford Co. (Brown 1989)
Brown, Joe: w 1989 Woodford Co. (CMW) [son of Billy Morgan Brown, above]
Brown, John: w 1980s Woodford Co. (Brown 1990) [son of Billy Morgan Brown, above]
Brown, John (IRE): b 1823 w 1850 Bourbon Co. (USC)
Brown, Kenneth: w 1989 Woodford Co. (Brown 1990) [grandson of Jesse Brown, Sr., nephew of Billy Morgan Brown, above]
Brown, Lonnie: w 1980s Woodford Co. (Brown 1990) [brother of Kenneth Brown, above]
Brown, William (IRE): b 1826 w 1850 Bourbon Co. (USC)
Brumfield?, James (KY): b 1807 w 1850 Jessamine Co. (USC)
Brumfield?, Thomas (KY): b 1835 w 1850 Jessamine Co. (USC) [son of James Brumfield, above]
Bunyan, William (KY): b 1799 w 1850 Mercer Co. (USC)
Burk, John: w 1851 Bourbon Co. (Clay 1846-1877)
Burk, Pat (IRE): b 1819 w 1870 Bourbon Co. (USC)
Burks, John (KY): b 1799 w 1850 Shelby Co. (USC)
Burnes, James (KY): b 1808 w 1850 Nelson Co. (USC)
Burnes, Stephen (KY): b 1820 w 1850 Nelson Co. (USC) [lives next door to James Burnes, above]
Burns, Edward (IRE): b 1800 w 1850 Woodford Co. (USC)

Burns, John (PA): b 1833 w 1880 Bourbon Co. (USC)
Burns?, Jonathan (IRE): b 1820 w 1870 Bourbon Co. (USC) [he and Jonathan Gregory listed with Terrence Kenny, below]
Butler, Edward (Canada): b 1833 w 1900 Scott Co. (USC)
Cain, John/John H. (IRE): b 1857/1859 w 1900 d 1930 Bourbon Co. (USC & Miller 1989b)
Cain, John H., Jr. (fa IRE): b c.1880 d c.1950 Bourbon Co. (Miller 1989b) [son of John Cain, above]
Cain, Thomas (fa IRE): b 1881 w 1900 Bourbon Co. (USC) [son of John Cain, above]
Cambell, James (IRE): b 1811 w 1880 Mercer Co. (USC)
Cameron, James (VA): b 1815 w 1850 Jessamine Co. (USC)
Canly?, Francis (KY): b 1825 w 1850 Franklin Co. (USC)
Cannon, Charles (IRE): b 1824 w 1860 Woodford Co. (USC)
Cans?, James (IRE): b 1823 w 1860 Bourbon Co. (USC)
Carew, Patrick (IRE): b 1808 w 1850 Harrison Co. (USC) [listed with Elijah Bailey, farmer, b Ky)
Carey, Daniel (fa KY): b 1848 w 1880 Mercer Co. (USC)
Carey, Michael (IRE): b 1810 w 1860 Scott Co. (USC)
Carmickle, Bowen (ancestors Scottish): w 1800s Mercer Co. (Carmickle 1990) [great-great-grandfather of Wayne Carmickle, below]
Carmickle, Jake (ancestors Scottish): b 1900 Mercer Co. (Carmickle 1990) [grandfather of Wayne Carmickle, below]
Carmickle, James (ancestors Scottish): b 1939 w 1990 Mercer Co. (Carmickle 1990) [father of Wayne Carmickle, below]
Carmickle, John William (ancestors Scottish): w 1800s Mercer Co. (Carmickle 1990) [great grandfather of Wayne Carmickle, below]
Carmickle, Wayne (ancestors Scottish): w 1973-1990 Mercer Co. (*Springfield Sun* 25 July 1990)
Carney? Canney?, Will (IRE): b 1820 w 1870 Bourbon Co. (USC)
Carpenter, Anderson (TN): b 1810 w 1850 Boyle Co. (USC)
Carpenter, James (KY): b 1814 w 1850 Shelby Co. (USC)
Carr, James (IRE): b 1830 w 1870 Bourbon Co. (USC)
Carr, John (IRE): b 1820 w 1870 Bourbon Co. (USC)
Carroll, James (VA): b 1788 w 1850 Jessamine Co. (USC) [listed with Goforth family, all b KY]
Carter, Joshua (KY): b 1817 w 1850 Anderson Co. (USC)
Carter, Patmo (IRE): b 1820 w 1860 (quarrier) Scott Co. (USC) [and family, listed with Thomas O'Connor and family, contractor, b IRE and 11 other Irish quarriers, some with families]
Case, Henry (KY): b 1805 w 1850 Mason Co. (USC)
Case? Call?, Moses (IRE): b 1807 w 1850 Bourbon Co. (USC)
Chambers, James (PA): b 1791 w 1850 Harrison Co. (USC)
Chancellor, Julias (KY): b 1798 w 1850, 1860 Fayette, Woodford Cos. (USC) [listed 1850 with Hugh Loney, grocer, b IRE]
Charity, Dominie (IRE): b 1939 w 1880 Mercer Co. (USC)
Cheek, Theo (KY): b 1835 w 1870 Mercer Co. (USC)
Cheek?. Thomas S. (fa KY): b 1836 w 1880 Woodford Co. (USC)
Cheek, William (fa KY): b 1831 w 1850 Shelby Co. (USC)
Childress, P. (KY): b 1811 w 1850 Nelson Co. (USC)

Clancy, Pat (IRE): b 1812 w 1860 Bourbon Co. (USC)
Clark?, Mathian (fa KY): b 1815 w 1880 Mercer Co. (USC)
Clary, Cornelius (IRE): b 1837 w 1860 Bourbon Co. (USC) [listed with J.S. Hill, grocer clerk, b MD]
Clem, Bill (par IRE): d 1988 Bourbon Co. (Caton 1988)
Click?, Mathias (KY): b 1815 w 1880 Mercer Co. (USC)
Cliff, Joseph (KY): b 1816 w 1870 Scott Co. (USC) [listed with John McGriffin, below]
Clutter?, William (KY): b 1810 w 1850 Bourbon Co. (USC)
Coady?, Thomas (IRE): b 1810 w 1860 Scott Co. (USC)
Coatney, Joshua (KY): b 1820 w 1880 Scott Co. (USC)
Cobert, James (KY): b 1823 w 1850 Mercer Co. (USC)
Colelazer, James (KY): b 1810 w 1850 Fayette Co. (USC)
Coleman, Henry (Germany): b 1808 w 1850 Scott Co. (USC) [listed with M. Johnson, occ not given, b KY]
Collins, Jacob (KY): b 1810 w 1850 Mason Co. (USC)
Collins, James (IRE): b 1810 w 1850 Clark Co. (USC)
Collins, John (IRE): b 1845 w 1880 Anderson Co. (USC)
Collopp, George (Prussia): b 1810 w 1870 Bourbon Co. (USC) [listed with Joseph Hall, farmer, b KY]
Comak, Palnche? (IRE): b 1840 w 1900 Bourbon Co. (USC)
Combs, Edward (KY): b 1828 w 1850 Clark Co. (USC)
Conley/Conely, Michael: w 1850 Bourbon Co. (Clay 1839-1853)
Conner, John (KY): b 1817 w 1850 Harrison Co. (USC)
Conway, Dan (IRE): b 1828 w 1860 (quarrier) Scott Co. (USC) [listed with Thomas O'Connor and family, contractor, b IRE, and 11 other Irish quarriers, some with families]
Conway, M. (IRE): b 1831 w 1880 Bourbon Co. (USC)
Cook, Jesse (KY): b 1798 w 1850 Franklin Co. (USC) [listed with Peter Tomlinson, ferryman, b KY]
Cook, John (KY): b 1802 w 1850 Shelby Co. (USC)
Cook, William (KY): b 1803 w 1850 Fayette Co. (USC)
Cook, William (KY): b 1795 w 1850 Harrison Co. (USC)
Coon, John (KY): b 1822 w 1870 Scott Co. (USC)
Corbat?, James (IRE): b 1816 w 1860 Bourbon Co. (USC) [listed with James Made? and fam, farm hand, b IRE]
Corn, Overton (fa KY): b 1821 w 1850 Mercer Co. (USC)
Cornelius, John (par KY): b 1834 w 1880 Scott Co. (USC)
Corns, John (KY): b 1830 w 1880 Scott Co. (USC)
Cosby, Thomas (KY): b 1823 w 1870 Mercer Co. (USC)
Coth?, Arther: w 1842 Mercer Co [Shaker Ledger Books 1839-1871]
Coughlin, Jerry (IRE): b 1822 w 1860 Scott Co. (USC)
Coughlin, Richard (IRE): b 1810 w 1860 Scott Co. (USC)
Cowdry, M.: w 1841 Mercer Co. [Shaker Ledger Books 1839-1871]
Cox, William (VA): b 1827 w 1850 Mercer Co. (USC) [listed with John Fitsimons, below]
Coyle, John (VA): b 1799/1802 w 1850, 1860, 1870 Woodford Co. (USC)
Coyle? Cozl?, John (VA): b 1804 w 1850 Nelson Co. (USC)
Coyle, John (KY): b 1848 w 1870 Woodford Co. (USC) [son of John Coyle b 1799, above]

Crauley, Harbard (KY): b 1828 w 1850 Fayette Co. (USC) [brother of Harrison and Washington Crauley, below]
Crauley, Harrison (KY): b 1814 w 1850 Fayette Co. (USC) [brother of Harbard and Washington Crauley, above and below]
Crauley, Washington (KY): b 1810 w 1850 Fayette Co. (USC) [brother of Harbard and Harrison Crauley, above]
Crawford, Henry (IRE): b 1808 w 1850 Mason Co. (USC) [lived in Canada and NJ before KY]
Crawford, Nat. (KY): b 1786 w 1860 Scott Co. (USC)
Creech, Eddie (ancestors IRE): w 1980s Franklin Co. (Creech 1990)
Creech, John S. (ancestors IRE): w 1980s Franklin Co. (Creech 1990)
Creely, Barney (IRE): b 1815 w 1850 Mercer Co. (USC) [listed with Harvie Cozine, brick mason, b KY]
Cross, Mike: w 1985 Fayette Co. (*Lexington Herald-Leader* 13 April 1985)
Crow, John (KY): b 1798 w 1850 Clark Co. (USC)
Crow/Crowe, Thomas (IRE): b 1834/1835 w 1860, 1880 Woodford Co. (USC)
Crumm?, William (KY): b 1824 w 1850 Boyle Co. (USC)
Crutcher, Ely (fa VA): b 1820 w 1850 Franklin Co. (USC)
Crutcher, Isaac (KY): b 1782 w 1850 Franklin Co. (USC)
Culp, Jonathan (KY): b 1798 b 1850 Harrison Co. (USC)
Cumming, James (IRE): b 1828 w 1860 (quarrier) Scott Co. (USC) [listed with Charly Linch and family, below, and 18 other Irish quarriers]
Cutsinger, Jerry: w 1989 Anderson Co. (Basham 1988)
D____, Mike (IRE): b 1820 w 1880 Bourbon Co. (USC) [listed with wife Mary, toll gate keeper, b IRE]
Daggan?, Michael (IRE): b 1823 w 1860 (quarrier) Scott Co. (USC) [listed with Thomas O'Connor and family, contractor, b IRE and 11 other Irish quarriers, some with families]
Dahad, Thomas (IRE): b 1829 w 1850 Fayette Co. (USC) [listed with Patrick O'Neal, below]
Daily, Jeremiah (IRE): b 1825 w 1870 Anderson Co. (USC)
Daisey, Mickel (g uncle IRE): b 1878 w 1910 Woodford Co. (USC)
Dale, Alfred (KY): b 1817 w 1850 Woodford Co. (USC)
Dalton, Lawrence (IRE): b 1810 w 1870 Scott Co. (USC)
Dargavell, William Scott (Scotland): b 1862 d 1934 Fayette Co. (Tuttle 1991)
Darnold, William (KY): b 1820 w 1850 Mason Co. (USC)
Daugherty, Samuel (fa IRE): b 1838 w 1900 Woodford Co. (USC)
Daugherty, William (KY): b 1829 w 1870 Bourbon Co. (USC)
Davidson, James (KY): b 1829 w 1850 Fayette Co. (USC)
Davis, B.P. (KY): b 1822 w 1870 Mercer Co. (USC)
Davis, Basil (KY): b 1823 w 1860 Mercer Co. (USC)
Davis, Elijah (fa KY): b 1833 w 1880 Woodford Co. (USC)
Davis, Henry (IN): b 1776 w 1850 Jessamine Co. (USC)
Davis, Jacob (MD): b 1781 w 1850 Harrison Co. (USC)
Davis, Joseph, (IRE): b 1815 w 1850 Jessamine Co. (USC)
Daviss, James (KY): b 1811 w 1850 Mercer Co. (USC)
Dawson, William (KY): w 1850 Anderson Co. (USC)
Dayley, Daniel (IRE): b 1815 w 1860 Bourbon Co. (USC)
Dayley, Michael (IRE): b 1842 w 1860 Bourbon Co. (USC)

Dearingor, Andrew (KY): b 1807 w 1860 Mercer Co. (USC)
Denney?, Dennis (IRE): b 1819 w 1850 Bourbon Co. (USC) [listed with Robert Hunt, turnpiker, b IRE]
Denney?, Michael (IRE): b 1821 w 1850 Bourbon Co. (USC) [listed with Robert Hunt, turnpiker, b IRE]
Dernasir?, Allen (KY): b 1828 w 1850 Shelby Co. (USC)
Derringer, Daniel (KY): b 1817 w 1850 Mercer Co. (USC)
Devlyn, John (IRE): b 1829 w 1860 Woodford Co. (USC)
Dewaree, Allen (KY): b 1827 w 1870 Mercer Co. (USC)
Dillon, John (IRE): b 1814 w 1850 Boyle Co. (USC) [listed with Pat Gallaher and Thomas McCown, below]
Dixon, William (KY): b 1838 w 1880 Scott Co. (USC)
Dolan, James: w 1890s Woodford Co. (Gormley 1987) [brother of John and Mike Dolan, below]
Dolan, John: w 1890s Woodford Co. (Gormley 1987) [brother of James and Mike Dolan, above and below]
Dolan, Mike: w 1890s Woodford Co. (Gormly 1987) [brother of James and John Dolan, above]
Donahoo, Garrett (IRE): b 1814 w 1850 Bourbon Co. (USC)
Donevan, Solomon (KY): b 1839 w 1860 Mercer Co. (USC)
Donley, John (IRE): b 1833 w 1860 Mercer Co. (USC)
Donnelly, Martin? (IRE): b 1833 w 1870 Bourbon Co. (USC)
Donnill, P.M. (IRE): b 1820 w 1850 Scott Co. (USC)
Donohue, William (IRE): b 1826 w 1860 Woodford Co. (USC) [listed with William L. Graddy, farmer, b KY]
Dooley, Timothy (IRE): b 1827 w 1880 Woodford Co. (USC)
Dotson, Charles (KY): b 1804 w 1850 Mason Co. (USC)
Dotson, Samuel M. (fa KY): b 1830 w 1850 Mason Co. (USC) [son of Charles Dotson, above]
Dougherty, James (IRE): b 1812 w 1850 Nicholas Co. (USC)
Douglas, James (IRE): b 1815 w 1850 Franklin Co. (USC)
Dowd, John (IRE): b 1836 w 1869-1873, 1880 Bourbon Co. (Clay 1854-1875; USC)
Dowden, Martin (IRE): b 1814 w 1850 (rock quarrier) Jessamine Co. (USC) [listed with Richard C. Graves, farmer & manufacturer, b KY]
Doyle, Dennis (KY): b 1858 w 1900 Scott Co. (USC)
Doyle, James (IRE): b 1825 w 1870, 1880 Anderson Co. (USC)
Doyle, John (IRE): b 1835 w 1870-1873 Bourbon Co. (Clay 1846-1877; USC) [could be same person as either or both John Doyle's, below]
Doyle, John (IRE): b 1837 w 1860 (quarrier) Scott Co. (USC) [listed with Thomas O'Connor and family, contractor, b IRE and 11 other Irish quarriers, some with families]
Doyle, John R. (IRE): b 1835 w 1910 Bourbon Co. (USC)
Doyle, Patrick (IRE): b 1814 w 1860 Mercer Co. (USC)
Dozier, Sidney (KY): b 1824 w 1850 Madison Co. (USC)
Duggins, Elijah (fa KY): b 1828 w 1850 Garrard Co. (USC)
Earlywine, William (KY): b 1850 w 1880 Bourbon Co. (USC)
Earnspiker, Ephraim (KY): b 1822 w 1850 Shelby Co. (USC)
Eaton, Jehu (KY): b 1802 w 1850 Anderson Co. (USC)

Eaves, James (KY): b 1811 w 1860 Woodford Co. (USC) [listed with Henry Morton, farmer, b KY)
Eddy, William (NY): b 1826 w 1870 Woodford Co. (USC)
Edwards, James (VA): b 1797 w 1850 Harrison Co. (USC)
Edwards, Thomas (KY): b 1810 w 1850 Harrison Co. (USC)
Egan, Mark (IRE): b 1816 w 1850 Franklin Co. (USC) [listed with Danric Lee, laborer, b IRE]
Eldridge, Lon: w 1990 Pulaski Co. (CMW)
Ellis, Benjamin (PA): b 1786 w 1850 Harrison Co. (USC)
Ency, Abraham (OH): b 1815 w 1860 Mercer Co. (USC)
Evans, Archer (PA): b 1822 w 1860 Scott Co. (USC)
Evans, Martin (IRE): b 1821 w 1850 Anderson Co. (USC)
Eve, William (KY): b 1835 w 1860 Woodford Co. (USC)
Fa____ll, James (IRE): b 1830 w 1870 Bourbon Co. (USC)
Fairweather, James (Scotland): b 1798 w 1850 Shelby Co. (USC) [listed with Samuel Dernasir?, blacksmith, b KY)
Fallen, Pat (IRE): b 1835 w 1870 Bourbon Co. (USC)
Fallice/Fallis, George (KY): b 1842 w 1860, 1880 Anderson Co. (USC) [son of William Fallice, below]
Fallice, William (KY): b 1815 w 1860 Anderson Co. (USC)
Fallis, Samuel (KY): b 1847 w 1880 Anderson Co. (USC)
Faoz, Charley (IRE): b 1805 w 1860 (quarrier) Scott Co. (USC) [lives with Charly Linch and family and 18 other Irish quarriers]
Faoz, Charley, Jr. (IRE): b 1836 w 1860 (quarrier) Scott Co. (USC) [son of Charley Faoz, above; lives with Charly Linch and family]
Faoz, John (IRE): b 1838 w 1860 (quarrier) Scott Co. (USC) [son of Charley Faoz, above; lives with Charly Linch and family]
Far____on, Farson?, J.B. (IRE): b 1820 w 1850 Nicholas Co. (USC)
Farrall, John (IRE): b 1830 w 1860 Mercer Co. (USC)
Farrall, Patrick (IRE): b 1810 w 1860 Mercer Co. (USC) [has son also named Patrick Farrall]
Farrell, James (IRE): b 1835 w 1880, 1900 Bourbon Co. (USC)
Ferguson, Michael (IRE): b 1830 w 1860 Woodford Co. (USC)
Ferguson, Mike (IRE): b 1835 w 1870 Woodford Co. (USC)
Ferlix, Ferdinand (Germany): b 1798 w 1850 Mason Co. (USC)
Ferlix, John (Germany): b 1833 w 1850 Mason Co. (USC) [son of Ferdinand Ferlix, above]
Ferrell, Mucajah (KY): b 1802 w 1850 Garrard Co. (USC)
Ferris?, Michael (IRE): b 1824 w 1850 Harrison Co. (USC)
Ficklin, Joseph: w 1845 Mercer Co. (Shaker Ledger Books 1839-1871)
Fields, William (KY): b 1814 w 1850 Mason Co. (USC) [listed with C.O. Whitescarver, farmer, b VA]
Finn, Andrew: w 1989 (*Lexington Herald-Leader* 26 September 1989, classified section)
Finn, Mike (IRE): b 1847 w 1880 Mercer Co. (USC)
Fitr? Fitres?, Jesse (KY): b 1845 w 1900 Scott Co. (USC)
Fitsimons, John (IRE): b 1817 w 1850 Mercer Co. (USC) [listed with William Cox, above]
Fitzpatreik?, James (par IRE): b 1866 w 1910 Scott Co. (USC)

Fitzpatrick, Dennis (IRE): b 1827 w 1850 Harrison Co. (USC)
Flarrity, Michael (IRE): b 1810 w 1860 Scott Co. (USC) [listedwith John O'Brian and family, below, and 6 other Irish stonemasons]
Fleming, Benjamin (KY): b 1845 w 1880 Anderson Co. (USC)
Fleming, James (IRE): b 1790 w 1860 Bourbon Co. (USC)
Fline?, Michael (IRE): b 1821 w 1860 Anderson Co. (USC) [he and Wm. Haley, below, listed with James D. Parker, farmer, b VA]
Florence, Nelsen (KY): b 1828 w 1850 Harrison Co. (USC)
Florence, Nicholas (PA): b 1814 w 1850 Harrison Co. (USC)
Florence?, William (KY): b 1819 w 1850 Harrison Co. (USC)
Flynn, John (KY): b 1842 w 1900 Mercer Co. (USC)
Flynn, John R. (IRE): b 1849 w 1880 Woodford Co. (USC)
Flynn, Maurice (fa IRE): b 1867 w 1900 Woodford Co. (USC)
Flynn, Michael (IRE): b 1820 w 1850 Franklin Co. (USC) [he and Patrick Oleonnes, below, listed with Mary Hampton b KY and other boarders?]
Flynn, Richard V. (par IRE): b 1875 w 1900 Mercer Co. (USC)
Follis, Samm? (KY): b 1835 w 1870 Anderson Co. (USC)
Fooley? Forley?, John (IRE): b 1810 w 1860 Woodford Co. (USC)
Forsten?, A.J.? (KY): b 1830 w 1860 Woodford Co. (USC)
Forston, R.J. (KY): b 1829 w 1870 Mercer Co. (USC) [probably same person as A.J. Forsten]
Foxworthy, Darius: w mid-1800s Scott Co. (Bevins 1984a)
Fralen, John (IRE): b 1835 w 1860 Scott Co. (USC) [lives with John O'Brian and family, below, and 6 other Irish stonemasons]
Franklin, James (SC): b 1805 w 1850 Jessamine Co. (USC)
Galbreath, George (KY): b 1806 w 1850 Mason Co. (USC)
Gallager, James (IRE): b 1835 w 1879 Woodford Co. (USC)
Gallaher, Pat (IRE): b 1825 w 1850 Boyle Co. (USC) [listed with John Dillon and Thomas McCown, above and below]
Galvin, Jerry (IRE): b 1810 w 1860 (quarrier) Scott Co. (USC) [listed with Thomas O'Connor and family, contractor, b IRE, and 11 other Irish quarriers, some with families]
Gannon, John (IRE): b 1830 w 1880 Bourbon Co. (USC)
Gardner, Andrew (MD): b 1791 w 1850 Harrison Co. (USC)
Garnett, Edward (KY): b 1808 w 1850 Harrison Co. (USC)
Garrett, Barry (IRE): b 1830 w 1860 (quarrier) Scott Co. (USC)
Garrison, Ted: w 1989 Mercer Co. (CMW)
Gary, David (KY): b 1848 w 1880 Mercer Co. (USC)
Gatewood, ? (KY): b 1868 w 1900 Scott Co. (USC)
George, William (KY): b 1820 w 1860 Woodford Co. (USC)
Gerry, Michael (IRE): b 1838 w 1860 (quarrier) Scott Co. (USC)
Gibson, John L. (KY): b 1881 w 1910 Mercer Co. (USC)
Giles, Lawrence (IRE): b 1821 w 1850 Woodford Co. (USC)
Givel?, Garven (IRE): b 1837 w 1860 (quarrier) Scott Co. (USC) [lives with Charly Linch and family, below, and 18 other Irish quarriers]
Go____gs, Sanford (KY): b 1815 w 1850 Franklin Co. (USC)
Goda?, Peter (IRE): b 1825 w 1860 Scott Co. (USC) [lives with Charly Linch and family, below, and 18 other Irish quarriers]
Goins, Samuel H. (KY): b 1819 w 1850 Garrard Co. (USC)

Goforth, Herbert (ancestors IRE): w mid-1900s Jessamine Co (Kelly 1989c) [taught Stanley Kelly, below]
Goforth, father of Herbert, above (ancestors IRE): w early 1900s Jessamine Co (Kelly 1989c)
Goforth, grandfather of Herbert, above (IRE): w late 1800s Jessamine Co. (Kelly 1989c)
Golden, Louis (VA): b 1812 w 1850 Scott Co (USC)
Goins, Joseph (KY): b 1815 w 1850 Franklin Co (USC)
Goren, Christopher (IRE): b 1826 w 1870 Bourbon Co (USC)
Gormley, Hugh (fa IRE): b late-1800s Woodford Co (Gormley 1987) [son of Michael Augustine Gormley, below]
Gormley, James Rogers (fa IRE): b 1885 d 1958 Woodford Co. (Gormley 1987) [son of Michael Augustine Gormley, below]
Gormley, John (fa IRE): b last ¼ 1800s Woodford Co. (Gormley 1987) [son of Michael Augustine Gormley, below]
Gormley, Joseph Pat (Bud) (fa IRE): b last ¼ 1800s Woodford Co. (Gormley 1987) [son of Michael Augustine Gormley, below]
Gormley, Michael (KY): b 1825 w 1860 Scott Co. (USC)
Gormley, Michael (IRE): b 1830 w 1860 Scott Co. (USC; Gormley 1990) [cousin of Michael Augustine Gormley, below]
Gormley, Michael A., Jr. (fa IRE): b late-1800s Woodford Co. (Gormley 1987) [son of Michael Augustine Gormley, above]
Gormley, Michael Augustine (IRE): b 1847 d 1922 Woodford Co. (Gormley 1987)
Gormley, Pat (IRE): b 1830 w 1870 Scott Co. (USC)
Gormley, Paul J. (grand fa IRE): b 1931 assisted as a youth 1940s Woodford Co. (Gormley 1987) [son of James Rogers Gormley, above]
Gormley, William (Willie) (fa IRE): b 1877 w 1910 Woodford Co. (USC; Gormley 1987) [son of Michael Augustine Gormley, below]
Gornly?, Thomas (IRE): b 1831 w 1860 Bourbon Co. (USC)
Grady, Michael (IRE): b 1852 w 1880 Woodford Co. (USC)
Gravitt, George (VA): b 1783 w 1850 Grant Co. (USC) [listed with James Hutton, farmer, b KY]
Grayson, Thomas (KY): b 1802 w 1850 Bracken Co. (USC)
Grayson, James (KY): b 1804 w 1850 Bracken Co. (USC) [lives two houses from Thomas Grayson, above]
Greeley, Barney (IRE): b 1815 w 1850 Mercer Co. (USC)
Green, Edward (KY): b 1790 w 1850 Clark Co. (USC)
Green, Mike (IRE): b 1833 w 1860 Bourbon Co. (USC) [listed with Pat Fallon, no occ, b IRE]
Gregery, Jonathan (IRE): b 1825 w 1870 Bourbon Co. (USC) [he and Jonathan Burns, above, listed with Terrence Kenny, below]
Gregory, George (IRE): b 1840 w 1870, 1900 Bourbon Co. (USC) [listed with B.B. Marsh, farmer, b KY]
Greiser, Ignas (Germany): b 1824 w 1850 Mason Co. (USC)
Gribbith, Charley (KY): w 1900 Scott Co. (USC)
Gribbon, Patrick (KY): b 1816 w 1850 Bourbon Co. (USC) [this is probably Patrick Griffin, below]
Griffin, John (KY): b 1862 w 1910 Scott Co. (USC)
Griffin, Patrick: w 1852/1853 Bourbon Co. (Clay Family Papers 1852, 1853)

Griffin, Robert: w 1869 Bourbon Co. (Clay 1854-1875)
Griffith, J.E. (KY): b 1837 w 1870 Scott Co. (USC)
Gunsales, A. (KY): b 1824 w 1860 Bourbon Co. (USC)
Gunsaules?, Theodore (KY): b 1805 w 1850 Harrison Co. (USC)
H____?, John (IRE): b 1825 w 1880 Anderson Co. (USC) [listed with brother-in-law, toll gate keeper, b IRE]
H?N?K?____h, Patrick (IRE): b 1825 w 1870 Bourbon Co. (USC)
H?N?K?____y, William (IRE): b 1843 w 1870 Bourbon Co. (USC)
Hagan, Michael (IRE): b 1832 w 1860 Scott Co. (USC) [brother of Stephen Hagan, below]
Hagan, Stephen (IRE): b 1830 w 1860 Scott Co. (USC) [brother of Michael Hagan, above]
Hagen, Alex (IRE): b 1838 w 1860 (quarrier) Scott Co. (USC) [lives with Charly Linch and family, below, and 18 other Irish quarriers]
Haggen? Haggin?, James (IRE): b 1840 w 1860 Mercer Co. (USC) [lives with Patrick Malary and Edward Bradling, below and above]
Halcomb, Hence: w 1971 Harlan Co. (*Cumberland Tri-City News* 24 Oct. 1971)
Haley, Dennis (IRE): b 1835 w 1860 (quarrier) Scott Co. (USC) [listed with Thomas O'Connor and family, contractor, b IRE and 11 other Irish quarriers, some with families]
Haley, William (IRE): b 1810 w 1860 Anderson Co. (USC) [he and Michael Fline, above, listed with James D. Parker, farmer, b VA]
Hall, David (KY): b 1825 w 1880 Woodford Co. (USC)
Hall, William (KY): b 1796 w 1850 Harrison Co. (USC)
Hall, William P. (KY): b 1821 w 1850 Nelson Co. (USC)
Halpern, Wm. (IRE): b 1819 w 1870 Scott Co. (USC)
Halpin, William (IRE): b 1823 w 1880 Scott Co. (USC)
Ham, Jacob (KY): b 1813 w 1850 Nicholas Co. (USC)
Hamilton, Thomas (VA): b 1770 "pauper" 1850 Harrison Co. (USC) [lives in poor-house]
Hand, Patrick (IRE): b 1805 w 1839 KY, 1850 Maury Co. TN (USC)
Handley, Michael (IRE): b 1826 w 1850 Franklin Co. (USC) [son of Morris Handley, below]
Handley, Morris (IRE): b 1800 w 1850 Franklin Co. (USC)
Handley, Thomas (IRE): b 1820 w 1850 Franklin Co. (USC) [son of Morris Handley, above]
Hanley, James (IRE): b 1832 w 1860 Scott Co. (USC)
Hanley, John (IRE): b 1828 w 1880 Anderson Co. (USC)
Hanlin?, Michael (IRE): b 1831 w 1860 Anderson Co. (USC) [listed with James Collins, grocer, b KY]
Hardin, George W. (KY): b 1816 w 1850 Harrison Co. (USC)
Hardin, James (KY): b 1821 w 1850 Garrard Co. (USC)
Hardwick, John (KY): b 1815 w 1850, 1860 Mercer Co. (USC)
Harkens, Thomas L. (KY): b 1834 w 1860 Bourbon Co. (USC) [listed as master stonemason, lives with George W. Davis and fam, master cabinet maker, b KY]
Harp, Austin: w mid-1900s Scott Co. (Harp 1988)
Harp, G.: w 1842 Mercer Co. (Shaker Ledger Books 1839-1871)
Harp, Paul: w 1989 Fayette Co. (Harp 1988) [son of Austin Harp, above]
Harp, S.: w 1842 Mercer Co. (Shaker Ledger Books 1839-1871)

Harris, Russell (KY): b 1810 w 1850 Garrard Co. (USC)
Harrison, Henry: w early 1900s Scott Co. (Bevins 1988)
Hartman, Tony: w 1900s Fayette Co. (McGregor 1988)
Hartnell, Daniel (IRE): b 1825 w 1860 Anderson Co. (USC) [listed with John Owens, laborer, b IRE]
Harvie, William (KY) b 1804 w 1850 Franklin Co. (USC)
Hawkins, Andrew (IRE): b 1820 w 1860 Woodford Co. (USC)
Hayden, Jonathan (KY): b 1810 w 1870 Bourbon Co. (USC)
Hayden, Jonathan, Jr. (KY): b 1844 w 1870 Bourbon Co. (USC) [son of Jonathan Hayden, above]
Hayes, Thomas (IRE): b 1814 w 1860 Scott Co. (USC) [listed with James Taylor and James Murray, below]
Hennary?, John (IRE): b 1833 w 1860 Bourbon Co. (USC) [listed with Thomas Gornly, above]
Henry, Bernard (fa IRE): b 1867 w 1880 Woodford Co. (USC) [son of Frank Henry, below]
Henry, Daniel (fa IRE): b 1864 w 1880 Woodford Co. (USC) [son of Frank Henry, below]
Henry, Edward (fa IRE): b 1842 w 1870 Mercer Co. (USC)
Henry, Frank (IRE): b 1820 w 1880 Woodford Co. (USC)
Henry, James (par IRE): b 1862 w 1880 Scott Co. (USC) [listed with his brother who was born in Ireland]
Henry, James (KY): b 1880 w 1910 Mercer Co. (USC)
Henry, John (IRE): b 1798 w 1850 Mercer Co. (USC)
Henry, John (IRE): b 1836 w 1870 Mercer Co. (USC)
Henry, Joseph (fa IRE): b 1852 w 1880 Woodford Co. (USC) [son of Frank Henry, above]
Henry, Patrick (fa IRE): b 1844 w 1880 Woodford Co. (USC) [son of Frank Henry, above]
Henry, Thomas (TN): b 1832 w 1850 Jessamine Co. (USC)
Hensley, Robert: w 1971 Harlan Co. (*Lexington Herald-Leader* 24 Oct. 1971)
Hewill, J.W. (KY): b 1837 w 1880 Scott Co. (USC)
Hickey, James (IRE): b 1820 w 1860 (quarrier) Scott Co. (USC) [listed with Garret Fitzgivens, stone cutter, b IRE, and 2 other Irish quarriers]
Hickey, Pat (IRE): b 1825 w 1880 Scott Co. (USC)
Hickey, R.E.: w 1980s Fayette Co. (*Lexington Herald-Leader* 13 Apr. 1985)
Hickey, Richard: w 1985 Fayette Co. (*Lexington Herald-Leader* 13 Apr. 1985) [son of R.E. Hickey, above]
Hickey, Wayne: w 1985 Fayette Co. (*Lexington Herald-Leader* 13 Apr. 1981) [son of R.E. Hickey, above]
Hifer, John (VA): b 1804 w 1850 Boone Co. (USC)
Hifer, John (VA): b 1804 w 1850 Boone Co. (USC)
Higgins, Morris (IRE): b 1814 w 1850 Scott Co. (USC) [listed with John Tanner, 70, millwright, b PA]
Higgins, Nimrod (KY): b 1820 w 1850 Mercer Co. (USC)
Higgs, Benny: w 1989 Woodford Co. (Brewer 1988)
Hill, John (KY): b 1852 w 1880 Scott Co. (USC)
Hill, Thomas (KY): b 1818 w 1850 Harrison Co. (USC)
Hill, William (NY): b 1815 w 1850 Franklin Co. (USC)

Hite, William (KY): b 1828 w 1860 Bourbon Co. (USC)
Ho____, Patrick (IRE): b 1787 w 1850 Boyle Co. (USC)
Ho____en, J. (IRE): b 1834 w 1880 Bourbon Co. (USC) [listed with J.?G. Crouch, farmer, b KY]
Hoar, Thomas (IRE): b 1834 w 1860 Bourbon Co. (USC)
Hockensmith, Charles: w 1988 Fayette Co. (Tate 1984)
Hoherty, ?: w 1871 Mercer Co. (Shaker Ledger Books 1839-1871)
Holbrook, Steve: w 1990 Scott Co. (Harp 1990)
Holdren, Samuel (IRE): b 1840 w 1880 Bourbon Co. (USC)
Holland, Jonathan (IRE): b 1842 w 1870 Bourbon Co. (USC)
Holleran, Michael (IRE): b 1829 w 1900 Bourbon Co. (USC)
Holleran, Simon (IRE): b 1837 Montgomery Co. (Holleran 1990)
Holleran, William (fa IRE): b 1884 Montgomery Co. (Holleran 1990) [son of Simon, above]
Honan, Austin (IRE): b 1845 w 1870 Bourbon Co. (USC)
Hore, J. (IRE): b 1815 w 1880 Bourbon Co. (USC)
Hore, J. (KY): b 1867 w 1880 Bourbon Co. (USC) [son of J. Hore, above]
Hornback, William (KY): b 1929 w 1860 Bourbon Co. (USC)
Hounegan, John (fa IRE): b 1838 w 1900 Bourbon Co. (USC)
Howarthy, William (KY): b 1830 w 1860 Bourbon Co. (USC) [lives next door to Jonathan Howerthy, below]
Howard, James (England): 1836 w 1900 Mercer Co. (USC)
Howe, David: w 1980s Fayette Co. (Tate 1984)
Howerthy, Jonathan (England): b 1830 w 1860 Bourbon Co. (USC)
Huff, Jack (KY): b 1799 w 1850 Mercer Co. (USC)
Huff, James: w early-1900s Mercer Co. (Rhodes 1989)
Huff, ______: w early-1900s Mercer Co. (Rhodes 1989) [father of James, above]
Huff, ______: w early-1900s Mercer Co. (Rhodes 1989) [brother of James, above]
Hunt, George Henry (par IRE): b 1862 d 1936 Nelson Co. (Hunt 1990) [son of Timothy Hunt, below]
Hunt, Jesse P. (KY): b 1816 w 1850 Clark Co. (USC) [son of Jonathan Hunt, farmer, b NC]
Hunt, Timothy (IRE): b 1830 d 1886 Nelson Co. (Hunt 1990)
Hurley, John (IRE): b 1800 w 1860 (quarrier) Scott Co. (USC) [listed with Garret Fitzgivens, stone cutter, b IRE, and 2 other Irish quarriers]
Husse?, Jan (IRE): b 1837 w 1860 (quarrier) Scott Co. (USC) [lives with Charly Linch and family, below, and 18 other Irish quarriers]
Isac, Owen (KY): b 1837 w 1880 Woodford Co. (USC)
Jacobs, Presley (KY): b 1820 w 1850 Shelby Co. (USC)
Jennen, John (IRE): b 1810 w 1860 Scott Co. (USC) [lives with Charly Linch and family, below, and 18 other Irish quarriers]
Jeter, Benjamin (KY): b 1818 w 1850 Fayette Co. (USC)
Jiles, Johnathan (KY): b 1820 w 1860 Mercer Co. (USC)
Johnson, Michael (IRE): b 1807 w 1860 Woodford Co. (USC)
Jones, Alen: w last part 1800s Grant Co. (Isham 1989) [son of Thomas P. Jones, below]
Jones, Daniel (PA): b 1788 w 1850, 1860 Bourbon Co. (USC)
Jones, George (VA): b 1796 w 1850 Harrison Co. (USC)
Jones, Henry (KY): b 1831 w 1870 Mercer Co. (USC)
Jones, James (KY): b 1805 w 1850 Mercer Co. (USC)

Jones, Owen? Ivan? (Wales): b 1810 w 1850 Woodford Co. (USC)
Jones, Thomas P.: b 1846 d 1930 Grant Co. (Isham 1989)
Jones, William (KY): b 1854 w 1880 Bourbon Co. (USC)
Joyce Andrew (IRE): b 1820 w 1850 Harrison Co. (USC)
Joyce, John (IRE): b 1836 w 1860 (quarrier) Scott Co. (USC) [and family, listed with Thomas O'Connor and family, contractor, b IRE, and 11 other Irish quarriers, some with families]
Joyce, Mike: w 1910-1920 Bourbon Co. (Letton 1989)
K?H?N?e____iff, Bart (IRE): b 1835 w 1870 Bourbon Co. (USC)
Kearney, Felix (IRE): b 1815 w 1860 Scott Co. (USC)
Kearney, John (IRE): b 1824 w 1860, 1900 Woodford Co. (USC)
Kearney, John (fa IRE): b 1856 w 1900 Woodford Co. (USC)
Keefe, James (IRE): b 1824 w 1860 Bourbon Co. (USC)
Keefe, John (IRE): b 1821 w 1860 Bourbon Co. (USC) [lives next door to James Keefe, above]
Keith, John (IRE): b 1822 w 1880 Woodford Co. (USC)
Keller, Morris (IRE): b 1850 w 1870 Scott Co. (USC) [listed with John Thompson, below]
Kelley, Michel (IRE): b 1830 w 1880 Bourbon Co. (USC)
Kelley, Patrick (IRE): b 1820 w 1880 Bourbon Co. (USC)
Kelly, Stanley: w 1989 Mercer Co. (CMW)
Kelly, Thomas (IRE): b 1810 w 1850 Hart Co. (USC) [listed with Patrick Crow, farmer, b IRE)
Kelly, Will (IRE): b 1825 w 1860 Bourbon Co. (USC)
Kennaday, Claborne (KY): b 1807 w 1850 Nicholas Co. (USC)
Kennady, Thomas (KY): b 1798 w 1850 Mason Co. (USC)
Kennedy, Bill: w 1867 Bourbon Co. (Clay 1854-1875)
Kenney, Patrick (IRE): b 1814 w 1850 Bourbon Co. (USC) [listed with Patrick Cooney, turnpiker, b IRE]
Kenny?, Terrence (IRE): b 1820 w 1870 Bourbon Co. (USC)
Kensley, John (IRE): b 1827 w 1860 Scott Co. (USC)
Kidd, Anderson (KY): b 1822 w 1850 Garrard Co. (USC)
Kills, James (IRE): b 1819 w 1850 Boone Co. (USC) [he and William Philips, below, listed with John Loar, bricklayer, b OH]
King: w 1869 Bourbon Co. (Clay 1854-1875)
King, Green B. (KY): b 1814 w 1850 Harrison Co. (USC)
King, John (KY): b 1850 w 1900 Mercer Co. (USC)
King, Seraphin (Germany): b 1818 w 1850 Mason Co. (USC)
Kolapp, George (Germany): b 1810 w 1860 Bourbon Co. (USC) [listed with Susan Smith and fam, no occ given, b KY]
Laferdy, Patrick (IRE): b 1830 w 1880 Scott Co (USC)
Laffin, M. (IRE): b 1815 w 1850 Scott Co. (USC) [listed with M. Crenshaw, female, b VA]
Lair, William (KY): b 1820 "convict" 1850 Boyle Co. (USC)
Lambert, Joseph (IRE): b 1820 w 1860 Woodford Co. (USC)
Land?, John (KY): b 1808 w 1850 Madison Co. (USC)
Lane, John (IRE): b 1791 w 1850 Woodford Co. (USC) [he and wife and other Irish laborers live in tavern of John Mitchell, farmer, b KY]

Lane, John (IRE): b 1789 w 1850 Franklin Co. (USC) [probably the same as John Lane, b 1791, counted twice]
Langhorn, Isaac (KY): b 1833 w 1850 Fayette Co. (USC)
L?arkin, Cassius C. (KY): b 1810 w 1850 Fayette Co. (USC) [listed with Henry Saxton, sign painter, b MS]
Laughlin, Pat (IRE): b 1812 w 1870 Bourbon Co. (USC)
Lauler, Thomas (KY): b 1815 w 1870 Bourbon Co. (USC)
Laullen?, Will (IRE): b 1795 w 1870 Bourbon Co. (USC)
Laws, George (KY): b 1830 w 1850 Scott Co. (USC) [he and Wm. M'Henney, below, listed with Junius R. Ward, farmer, b KY]
Lawson, John: w late-1800s Nicholas Co. (Waugh 1988)
Layton, William (KY): b 1792 w 1850 Garrard Co. (USC)
Leden, Sam (IRE): b 1810 w 1860 Scott Co. (USC) [lives with John O'Brian and family and 6 other Irish stonemasons]
Leonard, Edward (IRE): b 1800 w 1850 Franklin Co. (USC)
Lewis, Isaac (KY): b 1806 w 1860 Woodford Co. (USC)
Lewis, Legran? (KY): b 1836 w 1880 Woodford Co. (USC)
Lewis, Robert (KY): b 1841 w 1860 Woodford Co. (USC) [son of Isaac Lewis, above]
Lewis, Samuel (KY): b 1833 w 1860 Woodford Co. (USC) [son of Isaac Lewis, above]
Linch, Charly (IRE): b 1808 w 1860 (quarrier) Scott Co. (USC)
Linehan, George (IRE): b 1833 w 1880 Bourbon Co. (USC) [listed with wife, Annora?, toll gate keeper, b IRE]
Linn, Edward (IRE): b 1820 w 1860 Bourbon Co. (USC)
Linn, Kelly: b 1887 d 1956 Scott Co. [Bevins 1984a]
Liston, William (IRE): b 1840 w 1870 Mercer Co. (USC)
Litrall, Richard (KY): b 1825 w 1850 Clark Co. (USC)
Loar, Nathan (MD): b 1786 w 1850 Boone Co. (USC)
Logue, John (IRE): b 1828 w 1870 Mercer Co. (USC)
Logue, Terrance (IRE): b 1805/1812 w 1850, 1870, 1880 Mercer Co. (USC)
Long, John (fa IRE): b 1853 w 1880 Bourbon Co. (USC)
Long, R. (IRE): b 1823 w 1880 Bourbon Co. (USC)
Long, Richard (IRE): b 1832 w 1860 Bourbon Co. (USC) [listed with Alex Craig, farmer, b KY]
Lory, Henry (MD): b 1813 w 1850 Mason Co. (USC) [listed with Henry Shea, trader, b IRE]
Love, Thomas (IRE): b 1835 w 1860 Scott Co. (USC) [listed with Barney Bradley, above]
Lowe, Squire (KY): b 1812 w 1850 Nicholas Co. (USC)
Lowe, William (KY): b 1830 w 1850 Nicholas Co. (USC) [listed with Squire Lowe, above]
Lusby, George (KY): b 1821 w 1850 Scott Co. (USC)
Lyllis, Thomas (IRE): b 1805 w 1850 Scott Co. (USC) [he and John Newnan and H. Mescher, below, are listed with John H. Wolfe, confectioner, b Germany]
Lynch, James (IRE): b 1815 w 1850 Scott Co. (USC) [F. Wallseir, below, listed with Lynch]
Lyons, Joseph (KY): b 1832 w 1880 Mercer Co. (USC)
Lytle, George: w 1848-1851 d by 1855 Bourbon Co. (Clay Family Papers 1839-1873)
Ma?arn?, Arthur (IRE): b 1815 w 1860 Bourbon Co. (USC)

McBride, Daniel (IRE): b 1805 "lunatic" 1850 Fayette Co. (USC)
McBride, J.M.: w 1841 Mercer Co. (Shaker Ledger Books 1939-1871)
McCalla, Francis (IRE): b 1812 w 1850 Anderson Co. (USC)
McCan, Edward (IRE): b 1840 w 1860 Woodford Co. (USC)
McCarthy, A. (IRE): b 1835 w 1880 Bourbon Co. (USC)
McCarty, James (IRE): b 1820 w 1850 Scott Co. (USC) [2 laborers, b IRE, live with McCarty]
McCarty, Samuel (KY): b 1802 w 1850 Fayette Co. (USC)
McCasley, John (IRE): b 1830 w 1860 (quarrier) Scott Co. (USC) [listed with Charly Linch and family, above, and 18 other Irish quarriers]
McClane, Anderson (Scotland): b 1806? w 1860 (quarrier) Woodford Co. (USC) [listed with Robert Patton, herdsman, b Scotland]
McCoat, Hugh (IRE): b 1823 w 1870 Woodford Co. (USC)
McCoon, Barney (IRE): b 1835 w 1860 (quarrier) Scott Co. (USC) [lives with Charly Linch and family, below, and 18 other Irish quarriers]
McCort, Hugh (IRE): b 1835 w 1870 Woodford Co. (USC)
McCourt, Peter (IRE): b 1820 w 1860 Woodford Co. (USC)
McCown, Thomas (IRE): b 1814 w 1850 Boyle Co. (USC) [listed with John Dillon and Pat Gallaher, above]
McCristal, Mike (fa IRE): b 1842 w 1900 Mercer Co. (USC)
McCune?, George (KY): b 1825 w 1850 Jessamine Co. (USC) [listed with Eliza Crow, b KY]
McCune/McClure, M. (IRE): b 1815 w 1850 Mercer Co. (USC) [listed with a turnpiker, b IRE]
McDaniel, James W. (KY): b 1866 w 1910 Woodford Co. (USC)
McDermot? Dermot?, Thomas (IRE): b 1823 w 1860 Bourbon Co. (USC)
McDernott, W. (IRE): b 1820 w 1880 Bourbon Co. (USC) [listed with Jonathan Parmison? and fam, laborer, b KY]
McDonald, James (IRE): b 1815 w 1870 Scott Co. (USC)
McDonald, Jonathan (IRE): b 1813 w 1870 Woodford Co. (USC)
McDonald, P. (IRE): b 1820 w 1850 Scott Co. (USC)
McDonnell, T. (IRE): b 1833 w 1860 Bourbon Co. (USC)
McDowell, Henry (KY): b 1820 w 1870 Mercer Co. (USC)
McFarland, Nash. (KY): b 1800 w 1850 Oldham Co. (USC)
McFarland, Nashl. (KY): b 1829 w 1850 Oldham Co. (USC) [son of Nash. McFarland, above]
McGarty?, Robert (IRE): b 1840 w 1880 Mercer Co. (USC)
McGarvey, Frank: w 1900s (quarrier) Woodford Co. (Gormley 1987)
McGirk, James (IRE): b 1845 w 1880 Woodford Co. (USC)
McGirk, Roger (IRE): b 1838/1840 w 1870, 1880 Woodford Co. (USC)
McGirk, Thomas (fa IRE): b 1865 w 1880 Woodford Co. (USC) [son of Roger McGirk, above]
McGory, Patrick: w 1843 (quarrier) Bourbon Co. (Clay 1839-1853)
McGory, William: w 1844 (quarrier) Bourbon Co. (Clay 1839-1853)
McGrain, John (IRE): b 1828 w 1870, 1880 Bourbon Co. (USC)
McGriffin, John (IRE): b 1831 w 1870 Scott Co. (USC) [listed with Joseph Cliff, above]
McGurk, Charles (stepfa IRE): b 1850 w 1880 Anderson Co. (USC) [stepson of John Hanley, above]

M'Henney [McKenney?], William (IRE): b 1815 w 1850 Scott Co. (USC) [he and George Laws, above, listed with Junius R. Ward, farmer, b KY]
McHugh, J. (IRE): b 1815 w 1850 Scott Co. (USC) [he and E. Bornet, above, listed with Michael Hickey and 9 other laborers, all b IRE]
McInstall, John (IRE): b 1805 w 1850 Mercer Co. (USC)
McLighter, John (IRE): b 1836 w 1870 Scott Co. (USC)
McLighter, Lou (IRE): b 1838 w 1870 Scott Co. (USC)
McKee, Arch. (KY): b 1797 w 1850 Jessamine Co. (USC)
McKeever, John (IRE): b 1835 w 1870 Scott Co. (USC)
McKeune?, James (IRE): b 1808 w 1870 Bourbon Co. (USC)
McKinbie, Joseph (IRE): b 1809 w 1860 Bourbon Co. (USC) [listed with James Forman, farmer, b VA]
McKown?, John (IRE): b 1820 w 1860 Mercer Co. (USC)
McMichael, Mich (IRE): b 1818 w 1850 Mercer Co. (USC) [listed with Wm Phange, no occ, b NY]
McMillan, Samuel (VA): b 1807 w 1850 Nicholas Co. (USC)
McNamara, ? (IRE): b 1805 w 1860 Scott Co. (USC)
McNamara, John (IRE): b 1805 w 1860 Scott Co. (USC)
McQueen, Patrick (IRE): b 1820 w 1850 Bourbon Co. (USC)
McQuinn, Michael (IRE): b 1835 w 1860 Scott Co. (USC)
McRaiur? McRaiun?, John (IRE): b 1820 w 1860 Mercer Co. (USC)
McSherry, Philip (IRE): b 1829 w 1900 Bourbon Co. (USC)
McWilliams, James (IRE): b 1828 w 1880 Woodford Co. (USC)
Madagan, D.: w 1847 Mercer Co. (Shaker Ledger Books 1839-1871)
Maddeon, John (IRE): b 1825 w 1870 Bourbon Co. (USC) [listed with Nora Davis and fam, black, laundress]
Maginley, Henry (fa IRE): b 1870 w 1900 Bourbon Co. (USC) [son of Owen Maginley, below]
Maginley, James (fa IRE): b 1876 w 1900 Bourbon Co. (USC) [son of Owen Maginley, below]
Maginley, John (fa IRE): b 1864 w 1900 Bourbon Co. (USC) [son of Owen Maginley, below]
Maginley, Owen (IRE): b 1828 w 1900 Bourbon Co. (USC)
MaGloffin, Michael (IRE): b 1820 w 1860 Scott Co. (USC)
Magurk, Barney (IRE): b 1828 w 1850 Mercer Co. (USC)
Mahar, J. (IRE): b 1836 w 1880 Bourbon Co. (USC)
Mahoney, Dennis (IRE): b 1838 w 1860 Bourbon Co. (USC) [listed with Jonathan Conner & wife, no occ, b IRE]
Mahoney, Mich (IRE): b 1818 w 1850 Mercer Co. (USC) [listed with Daniel Miller, blacksmith, b Germany]
Mahoney, Michael (IRE): b 1815? w 1860 Woodford Co. (USC)
Malary, Patrick (IRE): b 1815 w 1860 Mercer Co. (USC) [James Haggin and Edward Bradling, above, also in household]
Malona, James (IRE): b 1836 w 1860 Bourbon Co. (USC) [listed with Brutus Clay, farmer, b KY]
Malone, Thomas: w 1850 Bourbon Co. (Clay 1839-1853)
Manning, John (NY): b 1803 w 1850 Nelson Co. (USC)
Mannion?, ? (MD): b 1798 w 1860 Scott Co. (USC)

Mapre?, Franklin (MD): b 1802 w 1850 Franklin Co. (USC)
Martin, D. (KY): b 1821 Shelby Co. (USC)
Martin, Jeremiah (England): b 1794/1804 w 1850, 1860 Scott Co. (USC)
Martin, John (KY): b 1813 w 1850 Harrison Co. (USC)
Martin, Robert (IRE): b 1802 "lunatic" 1850 Fayette Co. (USC)
Martin, Robert (KY): b 1822 w 1850 Clark Co. (USC)
Mason, George (KY): b 1815 w 1850 Fayette Co. (USC)
Mason, Henry (SC): b 1806 w 1850 Boyle Co. (USC)
Masterson, John (IRE): b 1810 w 1870 Bourbon Co. (USC) [listed with Stephen Hays, runs coffee house, b IRE]
Matherly, ?: w 1980s Bourbon Co. (Dawson 1988)
Matthews, Thomas M. (KY): b 1821 w 1850 Grant Co. (USC)
Mayhall, Thomas (IRE): b 1822 w 1850 Woodford Co. (USC) [listed with Oscar Pepper, farmer, b KY]
Mayho, M. (IRE): b 1803 w 1850 Boyle Co. (USC)
Mays, Cary? Casey? (KY): b 1820 w 1850 Jessamine Co. (USC) [listed with Samuel Guyn, farmer, b KY]
Mays, John D. (KY): b 1828 w 1850 Jessamine Co. (USC)
Meadows, John (VA): b 1802 w 1850 Harrison Co. (USC)
Medlin, Richard (VA): b 1800 w 1850 Nicholas Co. (USC)
Mehan, Patrick (IRE): b 1820 w 1860 Mercer Co. (USC)
Mehan, William (IRE): b 1844 w 1860 Mercer Co. (USC) [son of Patrick Mehan, above]
Mellory, Thomas (IRE): b 1823 w 1860 Scott Co. (USC)
Mernagh?, Jonathan (IRE): b 1810 w 1870 Bourbon Co. (USC)
Mescher, H. (Germany): b 1823 w 1850 Scott Co. (USC) [he and John Newnan and Thomas Lyllis are listed with John H. Wolfe, confectioner, b Germany]
Mesher, Henry (IRE): b 1831 w 1870 Scott Co. (USC)
Mickie, B. (IRE): b 1830 w 1880 Mercer Co. (USC)
Mickie, John (par IRE): b 1864 w 1880 Mercer Co. (USC) [son of B. Mickie, above]
Mickie, Thomas (IRE): b 1859 w 1880 Mercer Co. (USC) [son of B. Mickie, above]
Miles, David: w 1989 Bourbon Co. (Hinkle 1989)
Miles, David, Sr.: w 1989 Bourbon Co. (Hinkle 1989)
Miles, John (KY): b 1841 w 1870 Woodford Co. (USC) [listed with Gabriel Morton, black, farm hand, b KY]
Miles, Levi (KY): b 1813 w 1850 Jefferson Co. (USC)
Miller, William (KY): b 1785 "lunatic" 1850 Fayette Co. (USC)
Mitchell, Hawkins (KY): b 1826 w 1850 Woodford Co. (USC)
Mitchell, Nathaniel (KY): b 1821 w 1850 Franklin Co. (USC)
Mitchell, Rick (ancestors IRE): w 1988 Fayette Co. (Caton 1988) [nephew of Bill Clem, above]
Mitchell, William F. (KY): b 1828 w 1850 Woodford Co. (USC)
Mitcheltree, Joe: w 1970s Nicholas Co. (CMW)
Moore, Harvey (KY): b 1803 w 1850 Fayette Co. (USC)
Moore, Henry (KY): b 1817 w 1850 Fayette Co. (USC) [lives with Robert Moore, below]
Moore, Henry (KY): b 1815 w 1870, 1880 Anderson Co. (USC)
Moore, James (KY): b 1828 w 1850 Franklin Co. (USC)
Moore, Moses (KY): b 1822 w 1850 Fayette Co. (USC)

Moore, Robert (KY): b 1807 w 1950 Fayette Co. (USC)
Moore, Thomas (KY): b 1827 w 1850 Scott Co. (USC)
Moreland, William (KY): b 1800 w 1850 Shelby Co. (USC)
Morlin, Isaac (IRE): b 1814 w 1850 Fayette Co. (USC)
Morris, Edmund (IRE): b 1827 w 1850 Franklin Co. (USC)
Morris, William (KY): b 1812 w 1850 Harrison Co. (USC)
Morrison, Joseph (MD): b 1783 w 1850 Mason Co. (USC) [listed with John Lunsford, farmer, b KY]
Morrow, William A. (KY): b 1818 w 1850 Nicholas Co. (USC)
Mortin, Ebenezer (Delaware): b 1810 w 1850 Franklin Co. (USC)
Morton, Medley (KY): b 1830 w 1860 Woodford Co. (USC)
Mouru?, James (IRE): b 1831 w 1860 Scott Co. (USC)
Muddiman, E.C.: w 1800s Scott Co. (Bevins 1984a)
Muddiman, Ely (England): b 1856 w 1880 Scott Co. (USC)
Muddman, Edward (England): b 1851 w 1880 Scott Co. (USC)
Mullen?, Barzella (KY): b 1807 w 1850 Harrison Co. (USC)
Murphy, C.B. (VA): b 1819 w 1870 Mercer Co. (USC)
Murphy, Charles W. (VA): b 1813 w 1860 Mercer Co. (USC)
Murphy, Cornelius (IRE): b 1827 w 1850 Harrison Co. (USC)
Murphy, Pat (IRE): b 1820 w 1860 Bourbon Co. (USC)
Murphy, William (KY): b 1823 w 1870 Mercer Co. (USC)
Murpree, John (IRE): b 1822 w 1850 Fayette Co. (USC)
Murray, John (IRE): b 1820 w 1850 Jessamine Co. (USC)
Murrey?, James (IRE): b 1820 w 1860 Scott Co. (USC) [listed with Thomas Hayes and James Taylor, above and below]
Nevell, Michael (IRE): b 1810 w 1850 Scott Co. (USC)
Newman, Patrick (IRE): b 1831 w 1860 Woodford Co. (USC)
Newnan, John (IRE): b 1805 w 1850 Scott Co. (USC) [he and Thomas Lyllis and H. Mescher, above, are listed with John H. Wolfe, confectioner, b Germany]
Noell, Frank (KY): b 1837 w 1880 Mercer Co. (USC)
Nolan, James (IRE): b 1810 w 1870 Bourbon Co. (USC) [listed with Charles Huffman, farmer, b KY]
Nooc, James F. (KY): b 1874 w 1900 Woodford Co. (USC)
Nooman, Michael (IRE): b 1830 w 1860 Woodford Co. (USC)
Nooman, Patrick (IRE): b 1838 w 1860 Woodford Co. (USC) [brother of Michael Nooman, above]
Nugent, Thomas (IRE): b 1827 w 1870 Woodford Co. (USC)
O'Brian, John (IRE): b 1800 w 1860 Scott Co. (USC) [7 other Irish stonemasons live with him and his family]
O'Bryan, Patrick (IRE): b 1817 w 1850 Bourbon Co. (USC)
O'Connell, Mike (IRE): b 1830 w 1870 Bourbon Co. (USC)
O'Connell, Peter (IRE): b 1842 w 1880 Bourbon Co. (USC)
O'Connor, Michael (IRE): b 1834 w 1860 Scott Co. (USC)
Oder, Albert (KY): b 1815 w 1850 Garrard Co. (USC)
Oder, Joseph (VA): b 1791 w 1850 Garrard Co. (USC)
O'Hara, Michael (IRE): b 1846 w 1880 Anderson Co. (USC)
O'Hare/Ohare, Pat/Patrick (IRE): b 1822/26 w 1860, 1870, 1880 Woodford Co. (USC)
Oleonnes?, Patrick (IRE): b 1825 w 1850 Franklin Co. (USC) [he and Michael Flynn listed with Mary Hampton, and other boarders?, b KY]

O'Neal, Henry: w 1842 Mercer Co. (Shaker Ledger Books 1839-1871)
O'Neal, Owen (IRE): b 1816 w 1870 Bourbon Co. (USC)
Oneal, P. (IRE): b 1823 w 1860 Scott Co. (USC)
O'Neal, Patrick (IRE): b 1817 w 1850 Fayette Co. (USC) [Thomas Dahad, above, listed with him]
O'Neal, Patrick (IRE): b 1820 w 1860 Anderson Co. (USC)
O'Neal, Peter (IRE): b 1847 w 1870 Bourbon Co. (USC) [son of Owen O'Neal, above]
O'Neal, Thomas (IRE): b 1810 w 1850 Mason Co. (USC) [listed with Robert Grundy, Presb. minister and landowner, b KY]
Owens, Eldredge (KY): b 1820 w 1850 Mason Co. (USC)
Owens, Milton (KY): b 1823 w 1850 Clark Co. (USC)
Owens, Willis (PA): b 1801 w 1850 Mason Co. (USC)
Ownes, Alfred B. (KY): b 1793 w 1850 Mason Co. (USC)
Paris, John W. (KY): b 1914 w 1850 Nicholas Co. (USC)
Parish, William (KY): b 1820 w 1850 Clark Co. (USC)
Parker, E. (KY): b 1829 w 1880 Anderson Co. (USC)
Parris, Jonathan (KY): b 1822 w 1870 Bourbon Co. (USC)
Parrish, Greenberry (KY): b 1824 w 1850 Clark Co. (USC)
Parrish, John: w 1837 Bourbon Co. (Field 1837)
Parrish, John (VA): b 1788 w 1850 Madison Co. (USC)
Parrish, Jonathan W. (KY): b 1820 w 1880 Bourbon Co. (USC) [listed with Elijah Walker and fam, farm laborer, b KY]
Parrish, William (IRE): b 1812 w 1852 Clark Co. (Parrish 1836-1882)
Paterson, Thomas (KY): b 1825 w 1870 Anderson Co. (USC)
Patrick, ? (IRE): b 1820 w 1850 Jessamine Co. (USC)
Pearce, E. (KY): b 1816 w 1850 Shelby Co. (USC)
Penn, James (KY): b 1813 w 1850 Franklin Co. (USC)
Penter?, Jonathan (IRE): b 1846 w 1870 Bourbon Co. (USC)
Perkins, James O. (KY): b 1877 w 1910 Bourbon Co. (USC)
Philips, William (VA): b 1800 w 1850 Garrard Co. (USC)
Philips, William J. (NY): b 1802 w 1850 Boone Co. (USC) [he and James Kills, above, listed with John Loar, brick layer, b OH]
Phillips, ? (KY): b 1835 w 1880 Mercer Co. (USC)
Phillips, G. (KY): b 1824 w 1880 Mercer Co. (USC) [probably same as George Phillips, below]
Phillips, George (KY): b 1826 w 1860 Mercer Co. (USC)
Phillips, John (VA): b 1807 w 1850 Nelson Co. (USC)
Phillsst?, George (KY): b 1826 w 1870 Mercer Co. (USC)
Pine, Daniel (IRE): b 1830/1835 w 1860, 1870 Woodford Co. (USC)
Pinkston, Allen (KY): b 1811 w 1850 Madison Co. (USC)
Piniora?, John (IRE); b 1811 w 1860 Scott Co. (USC)
Piper, Peter (Germany): b 1814 w 1850 Franklin Co. (USC)
Plunkett, James (IRE): b 1827 w 1860 Scott Co. (USC)
Powell, David (KY): b 1798 w 1850 Mason Co. (USC)
Powers?, Mike (KY): b 1842 w 1870 Bourbon Co. (USC)
Prather, Gene: w 1988 Fayette Co. (Hockensmith 1989)
Prather, James (KY): b 1842 w 1880 Anderson Co. (UCS)
Pray [Fray?], Patrick (IRE): b 1830 w 1860 Mercer Co. (USC)
Prench?, Pat (IRE): b 1843 w 1879 Bourbon Co. (USC)

Price, John (KY): b 1867 w 1900 Woodford Co. (USC)
Quinn, Jimmy: w early-1900s Robertson Co. (Wilson 1989)
Quinn, Steve: w early-1900s Robertson Co. (Wilson 1989)
R___tner?, William (KY): b 1800 w 1850 Madison Co. (USC)
Ragan, Jerry (IRE): b 1832 w 1860 Scott Co. (USC)
Rankin, Robert (KY): b 1800 w 1850 Harrison Co. (USC)
Read, John (IRE): b 1813 w 1850 Harrison Co. (USC)
Reed, Kenneth: w 1980s Woodford Co. (Huskisson 1988)
Renan, John (Scotland): b 1838 w 1860 Anderson Co. (USC) [listed with John Owens, laborer, b IRE]
Renner, Wendell: w 1990 Rockcastle Co. (Wooley 1990)
Reynolds, Levi (KY): b 1819 w 1850 Garrard Co. (USC)
Rhine, Richard (IRE): b 1817 w 1850 Woodford Co. (USC) [he and Patrick Welch, below, listed with George J. Graddy, farmer, b KY]
Rial? Bial?, John (KY): b 1823 w 1850 Jessamine Co. (USC)
Rice, William B. (VA): c 1785 w 1850 Nelson Co. (USC)
Richardson, Isaac (KY): b 1827 w 1850 Fayette Co. (USC)
Richardson, John (KY): b 1825 w 1850 Fayette Co. (USC)
Richardson, Thomas (SC): b 1798 w 1850 Jessamine Co. (USC)
Richardson, Thomas (SC): b 1795 w 1860 Woodford Co. (USC) [probably same person as Thomas Richardson, b 1798]
Rickman, William (KY): b 1822 w 1850 Mercer Co. (USC) [listed with Simon Engleman, waggoner, b KY]
Riggs, Clifton: w 1970s Nicholas Co. (CMW)
Riley, Barna (IRE): b 1825 w 1860 Scott Co. (USC) [listed with John O'Brian and family, above, and 6 other Irish stonemasons]
Riley, John (IRE): b 1825 w 1860 Mercer Co. (USC)
Riley, Michael (IRE): b 1820 w 1860 (quarrier) Scott Co. (USC) [listed with Charly Linch and family, above, and 18 other Irish quarriers]
Riley, Thomas (IRE): b 1825 w 1860 Scott Co. (USC) [listed with John O'Brian and family, above, and 6 other Irish stonemasons]
Riller? Ritter?, George (KY): b 1814 w 1850 Shelby Co. (USC)
Roberts, H.W.: w 1855 Nicholas Co. (Metzger 1989)
Roberts, Henry (KY): b 1814 w 1850 Harrison Co. (USC)
Roby? Boby?, Elevan (KY): b 1815 w 1870 Scott Co. (USC)
Roby, Joseph (KY): b 1818 w 1870 Scott Co. (USC)
Rochester, Craig/Greg: w 1980s Fayette Co. (CMW; Tate 1984)
Rogers, Isaac (KY): b 1796 w 1860 Bourbon Co. (USC)
Rogers, James (IRE): b 1815 w 1850 Clark Co. (USC) [listed with James Nelson, farmer, b VA]
Rogers, Mike (IRE): b 1835 w 1860 Bourbon Co. (USC)
Ronan?, James (IRE): b 1810 w 1860 Bourbon Co. (USC)
Ronan, John (IRE): b 1814 w 1860 Bourbon Co. (USC) [lives next door to James Ronan, above]
Ross, William (KY): b 1802 w 1850 Nicholas Co. (USC)
Rowan, James (g par IRE): b 1865 w 1880 Anderson Co. (USC) [son of Joseph Rowan, below]
Rowan, Joseph (par IRE): b 1842 w 1880 Anderson Co. (USC)
Rule, Jonathan (KY): b 1840 w 1870 Bourbon Co. (USC)

Runyan, William (KY): b 1799 w 1850 Mercer Co. (USC) [lives at Pleasant Hill]
Rurnp?, John (England): b 1848 w 1910 Woodford Co. (USC)
Ryan, Con (IRE): b 1830 w 1870 Bourbon Co. (USC)
Ryan, Thomas (IRE): b 1811 w 1850 Scott Co. (USC) [he and 2 laborers, b IRE, listed with lawyer, b KY]
Saclone, George (KY): b 1857 w 1870 Woodford Co. (USC)
Sallee, Jacob B. (KY): b 1792 w 1850 Jessamine Co. (USC)
Sallir, John (KY): b 1861 w 1900 Mercer Co. (USC)
Sallir, Robert H. (KY): b 1880 w 1900 Mercer Co. (USC)
Sampaen?, John G. (KY): b 1884 w 1910 Bourbon Co. (USC)
Sanders, Isaac, Jr. (KY): b 1811 w 1860 Mercer Co. (USC)
Savage, Jerry (IRE): b 1832 w 1860 (quarrier) Scott Co. (USC) [listed with Thomas O'Connor and family, contractor, b IRE and 11 other Irish quarriers, some with families]
Sayer, Thomas (IRE): b 1798 w 1850 Mercer Co. (USC)
Scanland, Micky (KY): b 1847 w 1870 Mercer Co. (USC)
Schottsman, Christion (Germany): b 1825 w 1850 Mason Co. (USC)
Sear, James M. (KY): b 1824 w 1880 Mercer Co. (USC)
Searcy?, John P. (KY): b 1851 w 1880 Anderson Co. (USC)
Sears, John (wife IRE): b 1837 w 1850 Mercer Co. (USC) [Michael Sullivan, below, also in household]
Sears, P.: w 1846 Mercer Co. (Shaker Ledger Books 1839-1871)
See, George (KY): b 1861 w 1900, 1910 Bourbon Co. (USC)
Sencer?, Surcer?, Berry? (KY): b 1793 w 1850 Boone Co. (USC)
Senet?/Senate, Peter (IRE): b 1824/30 w 1870, 1880 Woodford Co. (USC)
Sexton, Benny: w 1988 Garrard Co. (CMW)
Shackelford, James (KY): b 1826 w 1860 Mercer Co. (USC)
Shaiky, Owen (IRE): b 1835 w 1860 Mercer Co. (USC) [probably Owen Sharkey, listed twice]
Shanks, Francis (KY): b 1816 w 1850 Clark Co. (USC)
Shannon, Mike (IRE): b 1837 w 1880, 1900 Bourbon Co. (USC)
Shannon?, Samuel (KY): b 1869 w 1910 (quarryman) Woodford Co. (USC)
Shanting, Barney (IRE): b 1830 w 1860 (quarrier) Scott Co. (USC) [listed with Charly Linch and family, above, and 18 other Irish quarriers]
Shanting, Michael (IRE): b 1845 w 1860 (quarrier) Scott Co. (USC) [listed with Charly Linch and family, above, and 18 other Irish quarriers]
Sharkey, Owen (IRE): b 1836 w 1860 Mercer Co. (USC)
Sharp, Ezekiel (TN): b 1820 w 1850 Jessamine Co. (USC) [he and family listed with J.C. Coleman, farmer, b KY]
Shay, Dennis (IRE): b 1806 w 1860 (quarrier) Scott Co. (USC) [listed with Thomas O'Connor and family, contractor, b IRE, and 11 other Irish quarriers, some with families]
Shay, James (IRE): b 1843 w 1860 (quarrier) Scott Co. (USC) [probable son of Dennis Shay, listed with Thomas O'Connor and family, contractor, b IRE, and 11 other Irish quarriers, some with families]
Shea, Jonathan (IRE): b 1810 w 1880 Bourbon Co. (USC)
Shehan, Dennis (IRE): b 1820 w 1860 Woodford Co. (USC)
Shehan, Mike (IRE): b 1832 w 1880 Bourbon Co. (USC)
Shely, James (IRE): b 1805 w 1870 Woodford Co. (USC)

Shepherd, Alfred: w 1971 Harlan Co. (*Lexington Herald-Leader* 24 Oct. 1971)
Shields, Hugh (KY): b 1815 w 1850 Jessamine Co. (USC)
Shields, Hugh (KY): b 1813 w 1860 Woodford Co. (USC) [listed with Jonathan Shields, farmer, b KY]
Shields, Hugh (IRE): b 1817 w 1860 Woodford Co. (USC) [listed with D.J. Williams, Jr., farmer, b KY]
Shipley, Noah (KY): b 1811 w 1850 Mason Co. (USC) [listed as farmer and stonemason]
Shoa, James (IRE): b 1835 w 1860 Scott Co. (USC) [listed with John O'Brian, above, and family and 6 other Irish stonemasons]
Shopley, Samuel (MD): b 1810 w 1870 Bourbon Co. (USC) [listed with Maxwell M?W?inter, carpenter, b VA]
Shull, Martin (Prussia): b 1846 w 1880 Woodford Co. (USC)
Simpson, Clarence (KY): b 1883 w 1900 Woodford Co. (USC)
Sims, John: w 1989 Washington Co. (M. Smith 1989)
Sinate?, Peter (IRE): b 1822 w 1880 Woodford Co. (USC)
Smart, Daniel? (KY): b 1820 w 1880 Mercer Co. (USC)
Smart, James P. (KY): b 1815/17 w 1860, 1870 Mercer Co. (USC)
Smith, Charles B. (KY): b 1817 w 1850 Mason Co. (USC)
Smith, John (IRE): b 1807 w 1860 (quarrier) Scott Co. (USC) [lives with Charly Linch and family, above, and 18 other Irish quarriers]
Smith, John (IRE): b 1810 w 1860 Scott Co. (USC) [lives with John O'Brian and family, above, and 6 other Irish stonemasons]
Smith, John (KY): b 1828 w 1870 Scott Co. (USC)
Smith, Markes (KY): b 1812 w 1850 Jessamine Co. (USC)
Smith, Marvin T., Jr.: w 1980s Washington Co. (M. Smith 1989)
Smith, Napolean B. (KY): b 1824 w 1850 Mason Co. (USC) [son of Ruben B. Smith, below]
Smith, Richard: b 1901 d 1980 Mercer Co. (W.B. Smith 1989)
Smith, Ruben B. (VA): b 1790 w 1850 Mason Co. (USC)
Smith, Simeon (Connecticut): b 1760 "blind" 1850 Jefferson Co. (USC)
Smith, William Bradley: w 1980s Mercer Co. (Traynor 1989)
Smith, ______: w 1900s Mercer Co. (W.B. Smith 1989) [father of William Bradley]
Smither, Fantley: w 1980s (CMW; Tate 1984)
Smithy, Richard (KY): b 1785 w 1850 Mercer Co. (USC)
Smithy/Smithey, William (KY): b 1814 w 1850, 1860 Mercer Co. (USC)
Soper, Tom (ancestors English): w 1982 Bourbon Co. (T. Soper 1988)
Sorrell, J. (KY): b 1824 w 1850 Nelson Co. (USC) [son of Richard Sorrell, above]
Sorrell, Richard (VA): b 1795 w 1850 Nelson Co. (USC)
Sprig, James (IRE): b 1801 w 1860 Bourbon Co. (USC)
Srackk, Edward (IRE): b 1832 w 1850 Bourbon Co. (USC) [listed with Robert Hunt, turnpiker, b IRE]
Stack, Morris (IRE): b 1820 w 1860 (quarrier) Scott Co. (USC) [listed with Garret Fitzgivens, stone cutter, b IRE, and 2 other Irish quarriers]
Staley, John (KY): b 1834/1836 w 1910 d 1922 Scott Co. (USC; N. Conner 1989)
Stephens, Hadden (KY): b 1818 w 1850 Fayette Co. (USC)
Stephens, Nelson (KY): b 1813 w 1880 Bourbon Co. (USC)
Stevenson, James H. (NJ): b 1808 w 1850 Woodford Co. (USC)

Stewart, John (PA): b 1782 w 1850 Scott Co. (USC)
Stewart, Larry: w 1980s Clark Co. (Venable 1989)
Stillfield, John (KY): b 1812 w 1850 Fayette Co. (USC)
Stipes, Fedrick? (KY): b 1815 w 1880 Anderson Co. (USC)
Stipes, Kenny? (KY): b 1849 w 1870 Anderson Co. (USC)
Stuart, David (PA): b 1781 w 1860 Woodford Co. (USC)
Stuart, William (KY): b 1811 w 1850 Harrison Co. (USC)
Sullivan, Daniel (IRE): b 1828 w 1860 (quarrier) Scott Co. (USC) [listed with Thomas O'Connor and family, contractor, b IRE, and 11 other Irish quarriers, some with families]
Sullivan, Mathew (IRE): b 1815 w 1850 Woodford Co. (USC)
Sullivan, Michael (IRE): b 1832 w 1860 Mercer Co. (USC) [lives with John Sears, above]
Swaing?, David (IRE): b 1843 w 1870 Scott Co. (USC)
Tarpy, Auralaky? (IRE): b 1803 w 1870 Anderson Co. (USC)
Tatum, Samuel (VA): b 1805 w 1850 Garrard Co. (USC)
Tatum, Thomas (KY): b 1828 w 1850 Garrard Co. (USC)
Taylor, James (IRE): b 1824 w 1860 Scott Co. (USC) [lives with Thomas Hayes and James Murrey, above]
Taylor, John (KY): b 1842 w 1900 Bourbon Co. (USC)
Temple/Temples, Richard (KY): b 1816 w 1870 1880 Anderson Co. (USC)
Temples, Richard (NC): b 1785 w 1850 Anderson Co. (USC)
Terrin, Peter: w 1836, 1837 Bourbon Co. (Clay 1828-1846)
Thompson, Alender (KY): b 1805 w 1850 Fayette Co. (USC)
Thompson, Duncan (IRE): w mid-1800s Frankln Co. (T. Clark 1990)
Thompson, H. (KY): b 1868 w 1900 Woodford Co. (USC)
Thompson, John (IRE): b 1838 w 1870 Scott Co. (USC) [listed with Morris Keller, above]
Thompson, Joseph (KY): b 1830 w 1880 Woodford Co. (USC)
Thompson, Thomas (KY): b 1830 w 1850 Fayette Co. (USC) [son of Alender Thompson, above]
Thornton, Francis: w 1839-1850 Bourbon Co. (Clay Family Papers 1839-1850)
Tipton, Harvey: w 1989 Bourbon Co. (Clay 1990)
Toadvine, Fletcher (KY): b 1823 w 1850 Harrison Co. (USC)
Toadvine, Hiram (KY): b 1827 w 1850 Harrison Co. (USC)
Toadvine, McKendrie (KY): b 1828 w 1850 Harrison Co. (USC)
Todd, Baxter (KY): b 1814 w 1850 Madison Co. (USC)
Toles, Jonathan (KY): b 1814 w 1860 Mercer Co. (USC)
Toll, Jonathan (KY): b 1815 w 1850 Fayette Co. (USC)
Tolle, William T. (KY): b 1803 w 1850 Mason Co. (USC)
Toolen, Patrick (fa IRE): b 1860 w 1900 Bourbon Co. (USC)
Torpey, John (IRE): b 1806 w 1870 Woodford Co. (USC)
Tribble, Austin (KY): b 1895 w 1850 Clark Co. (USC)
Trower, Ervin? (KY): b 1830 w 1860 Mercer Co. (USC)
True, Thomas (KY): b 1834 w 1870 Bourbon Co. (USC)
Truitt, Comfort (Delaware): b 1766 w 1850 Scott Co. (USC)
Tucker, John W. (KY): b 1825 w 1850 Scott Co. (USC)
Tully, Thomas (IRE): b 1823 w 1870 Woodford Co. (USC)
Twing?, John (IRE): b 1835 w 1860 Scott Co. (USC)

Usher, Marcus (IRE): b 1815 w 1850 Mason Co. (USC) [listed with Hugh McCollough, tavern keeper, b IRE]
Vail, Morris, (KY): b 1818 w 1860 Scott Co. (USC)
VanArsdell, Peter (ancestors Dutch): 1800s Mercer Co. (Keightley 1989)
Vance, Jacob (VA): b 1780 w 1850 Garrard Co. (USC)
Vance, Roy: w early 1900s Scott Co. (Bevins 1988)
Vaughn, A.J. (KY): b 1828 w 1860 Mercer Co. (USC)
Vaughn, Allin (KY): b 1770 w 1850 Mercer Co. (USC)
Vinson, Andrew: w 1836, 1837 Bourbon Co. (Clay 1828-1846)
Wallace, Frederick (Germany): b 1805 w 1860 Scott Co. (USC)
Wallace, John (IRE): b 1824 w 1870 Anderson Co. (USC)
Wallseir?, F. (Germany): b 1810 w 1850 Scott Co. (USC) [listed with James Lynch, above]
Walsh, Tim (IRE): b 1813 w 1870 Bourbon Co. (USC)
Ward, Pat (IRE): b 1805 w 1870 Bourbon Co. (USC)
Warford, Ben Fountain (ancestors Irish): b 1864 d 1935 Franklin Co. (Worford 1988) [son of John Warford, Sr., below]
Warford, Brian (ancestors Irish): b c. 1860s Anderson Co. (Worford 1988) [son of John Warford, Sr., below]
Warford, John, Jr. (ancestors Irish): b c. 1860s Anderson Co. (Worford 1988) [son of John Warford, Sr., below]
Warford, John, Sr. (ancestors Irish): w mid-1800s Franklin Co. (Worford 1988)
Warford, Porter (ancestors Irish): b c. 1860s Anderson Co. (Worford 1988) [son of John Warford, Sr., above]
Warford, Tom (ancestors Irish): b c. 1860s Anderson Co. (Worford 1988) [son of John Warford, Sr., above]
Waters, Patrick (IRE): b 1840 w 1880 Woodford Co. (USC)
Watson, George M. (KY): b 1858 w 1900 Scott Co. (USC)
Watson, William H. (KY): b 1832 w 1850 Jessamine Co. (USC)
Watt/Watts, John (KY): b 1860 w 1910 Mercer Co. (USC)
Watts, George (IRE): b 1782 w 1850 Mason Co. (USC)
Watts, George A. (fa IRE): b 1830 w 1850 Mason Co. (USC) [son of George Watts, above]
Watts, John A. (fa IRE): b 1826 w 1850 Mason Co. (USC) [son of George Watts, above]
Watts, Robert A. (fa IRE): b 1829 w 1850 Mason Co. (USC) [son of George Watts, above]
Watts, Samuel W. (fa IRE): b 1832 w 1850 Mason Co. (USC) [son of George Watts, above]
Watts, William J. (fa IRE): b 1834 w 1850 Mason Co. (USC) [son of George Watts, above]
Watts, Willis (KY): b 1832 w 1870 Anderson Co. (USC)
Waugh, Bobby (ancestors Scottish): b 1938 Scott Co. (B. Waugh 1989) [son of Ernest Waugh, Sr., brother of Ernest Waugh, Jr., below]
Waugh, Darren (ancestors Scottish): b 1966 w 1980s Scott Co. (*Lexington Leader* 20 July 1982; B. Waugh 1989) [son of Bobby Waugh, above]
Waugh, Dennis (ancestors Scottish): b 1968 w 1989 Scott Co. (B. Waugh 1989) [son of Bobby Waugh]
Waugh, Ernest, Jr. (ancestors Scottish): b 1932 Bourbon Co. (L. Waugh 1989) [son of Ernest Waugh, Sr., below; brother of Bobby Waugh, above]

Waugh, Ernest, Sr. (ancestors Scottish): b 1905 d 1983 Bourbon Co. (Stallons 1976; B. Waugh 1989; L. Waugh 1989) [son of George Harrison Waugh, below]
Waugh, George Harrison (ancestors Scottish): b 1870 w 1870 d 1944 Nicholas Co (B. Waugh 1989) [father of Ernest Waugh, Sr., above]
Waugh, Larry (ancestors Scottish): b 1959 w 1989 Bourbon Co. (L. Waugh 1989) [son of Ernest Waugh, Jr., above]
Weaver, Jacob (Prussia): b 1812 w 1850 Boyle Co. (USC)
Webber, Michel (France): b 1814 w 1870 Scott Co. (USC)
Welch, Patrick (IRE): b 1825 w 1850 Woodford Co. (USC) [he and Richard Rhine listed with George J. Graddy, farmer, b KY]
Weloyner, Lewis (NC): b 1818 w 1850 Boyle Co. (USC)
Welsh, Jonathan (IRE): b 1801 w 1850 Bourbon Co. (USC) [listed with Rebecca Thomas, no occ given, b KY]
Welsh, Michael (IRE): b 1819 w 1850, 1860 Bourbon and Scott Cos. (USC)
Welsh, Thomas (IRE): b 1832 w 1860 (quarrier) Scott Co. (USC) [and family, listed with Thomas O'Connor and family, contractor, b IRE, and 11 other Irish quarriers, some with families]
West, BrightBerry? (KY): b 1832 w 1850 Madison Co. (USC) [son of John West, below]
West, John (KY): b 1807 w 1850 Madison Co. (USC)
West, Richard (KY): b 1804 w 1850 Madison Co. (USC)
West, Richard (KY): b 1830 w 1850 Madison Co. (USC) [son of John West, above]
Whaley, Hiram (KY): b 1810 w 1850 Nicholas Co. (USC)
Wickersham, Benjamin (KY): b 1795 w 1850 Mercer Co. (USC) [listed as brickmason in 1860]
Wiggins, Pat (par IRE): b 1844 w 1880 Mercer Co. (USC)
Wigham J.J. (KY): b 1860 w 1900 Mercer Co. (USC)
Wilbur?, John (KY): b 1848 w 1880 Anderson Co. (USC)
Williams, Ben: w early-1900s Robertson Co. (Wilson 1989)
Williams?, Charles? (KY): b 1853 w 1880 Anderson Co. (USC)
Williams, Ebenezer (KY): b 1809 w 1850 Nicholas Co. (USC)
Williams, Isaac (VA): b 1801 w 1850, 1860, 1870 Anderson Co. (USC)
Williams, Simeon (KY): b 1828 w 1850, 1860 Anderson Co. (USC)
Williamson, Alfred (KY): b 1820 w 1850 Boone Co. (USC) [son of Thomas Williamson, below]
Williamson, Thomas (KY): b 1796 w 1850 Boone Co. (USC)
Wilson, Jonas (KY): b 1814 w 1850 Fayette Co. (USC)
Wilson, Stephen (VA): b 1789 w 1850 Harrison Co. (USC)
Wilson, Valentine (KY): b 1823 w 1850 Jessamine Co. (USC)
Witt, Malcolm: w 1990 Woodford Co. (Wooley 1990)
Wolfe, Everett (KY): w 1900s Pendleton Co. (Wolfe 1990)
Wolfe, Robert: w 1980s Clark Co. (Venable 1989) [son of Everett Wolfe, above]
Woods, Edmond H. (fa IRE): b 1868 d 1905 Bourbon Co. (J. Miller 1989b) [son of Edmond P. Woods, below]
Woods, Edmond P. (IRE): b 1820 w 1870 d 1905 Bourbon Co. (USC; J. Miller 1989b)
Woods, Edward (IRE): b 1850 w 1900 Bourbon Co. (USC)
Woods, Jack/John (fa IRE): b 1860 w 1900 d 1917 Bourbon Co. (USC; Miller 1989b) [son of Edmond P. Woods, above]
Woods, Michael/Mike (fa IRE): b 1865/1870 w 1900 d 1943 Bourbon Co. (USC; Miller 1989b) [son of Edmond P. Woods, above]

Woods, Thomas Patrick "Kirby" (fa IRE): b 1855 d 1922 Bourbon Co. (Miller 1989b) [son of Edmond P. Woods, above]
Wooldridge, George H. (VA): b 1804 w 1860 (rock blaster) Mercer Co. (USC)
Worford, Charles (ancestors Irish): b 1910 Anderson Co. (Worford 1988) [son of Ben Fountain Warford, above]
Workman, Burel: w 1970s Bath Co. (CMW)
Yates: w 1837 Bourbon Co. (Clay 1828-1846)
Yeast, J.A. (KY): b 1817 w 1860 Mercer Co. (USC)
Yowel, Albert (KY): b 1837 w 1900 Woodford Co. (USC)
Yowel, Berry (KY): b 1842 w 1880 Woodford Co. (USC) [son of Connor Yowel, below]
Yowel, Connor (KY): b 1809 w 1880 Woodford Co. (USC)
Yress?, Wm. (KY): b 1810 w 1860 Scott Co. (USC)

Race Unknown

Barber, Johnny: w 1967 Scott Co. (Williamson 1967)
Gentry, Australia: w 1960s Fayette Co. (CMW)
Lee, Herbie: w 1980s Anderson Co. (Tate 1984)
Young, Tollie: w 1967 Bourbon Co. (Williamson 1967)

APPENDIX 2

Preservation Groups in the United Kingdom

For those interested in learning more about the efforts of groups active in rock fence preservation in Great Britain, their addresses are as follows:

Agricultural Training Board. Summit House, Glebe Way, West Wickham, Kent BR4 0RF.

Association for the Protection of Rural Scotland. 14a Napier Road, Edinburgh EH10 5AY.

British Trust for Conservation Volunteers. 36 St. Mary's Street, Wallingford, Oxon OX10 0EU.

Civic Trust. 17 Carlton House Terrace, London SW1Y 5AW.

Community Service Volunteers. 237 Pentonville Road, London N1 9NJ.

Conservation Trust. George Palmer Site, Northumberland Avenue, Reading RG2 7PW.

Council for British Archaeology. 112 Kennington Road, London SE11 6RE.

Council for the Protection of Rural England. 4 Hobart Place, London SW1W 0HY.

Country Landowner's Association. 16 Belgrave Square, London SW1X 8PQ.

Countryside Commission of England and Wales. John Dower House, Crescent Place, Cheltenham, Glos GL50 3RA.

Countryside Commission for Scotland. Battleby, Redgorton, Perth PH1 3EW.

Dry Stone Walling Association. YFC Centre, National Agricultural Centre, Kenilworth, Warwickshire CV8 2LG. (This group was especially helpful in providing valuable information on fence preservation.)

Central Scotland Branch, D.S.W.A. 8 Kingsway West, Dundee.

The Environment Council. 80 York Way, London WC2H 9HJ.

Field Studies Council. Preston Montford, Montford Bridge, Shrewsbury SY4 1HW.

The Friends of the Lake District. Gowan Knott, Kendal Road, Staveley, Kendal, Cumbria LA8 9LP.

Landscape Institute. 12 Carlton House Terrace, London SW1Y 5AH.

The National Trust. 36 Queen Anne's Gate, London SW1H 9AS.

The Open Spaces Society. 25a Bell Street, Henley-on-Thames, Oxon RG9 2BA.

Rural Development Commission. 141 Castle Street, Salisbury, Wiltshire SP1 3TP.

Scottish Conservation Projects Trust. Balallan House, 24 Allan Park, Stirling FK8 2QG.

Scottish Landowners' Federation. 18 Abercromby Place, Edinburgh EH3 6TY.

Stewartry Dry Stone Walling Committee. Cally Estate Office, Gatehouse of Fleet, Dumfries-Shire.

Stone Federation. 82 New Cavendish Street, London W1M 8AD.

Town and Country Planning Association. 17 Carlton House Terrace, London SW1Y 5AS.

Glossary

This glossary has two purposes. First, the definitions explain subtle but important details and variations in building techniques. Second, the cognate terms and their sources illustrate commonalities between terms used by historic and contemporary masons in Kentucky and the British Isles. This linguistic evidence supplements the historical record and reinforces our interpretation that Kentucky's rock fences were not local creations by antebellum bondsmen but were the product of a diffusion process that originated in Scotland, Ireland, and northern England.

The place where a term was or is used is followed by the name of the source in which the term was used or its use was documented and the source's date. Dates in brackets indicate dates of terminology use earlier than the source dates. We did not attempt to list all the terms used by every Kentucky fence mason but compiled this list from field notes of terms that arose incidentally in conversations and from English, Scottish, Irish, and American correspondence and publications. Starred terms (*) are defined in this glossary. Terms listed as "common usage" in Kentucky are from field notes; terms listed as "common usage" in Scotland, northern England, and the United Kingdom are from Tufnell 1990 and Dry Stone Walling Association 1989.

ABUTMENT: *See* pillar.

ANIMAL GAP: An opening in the lower part of the fence allowing small animals to go from one enclosure to another. Synonyms: **creep-hole** (England, Garner 1984); **cripple** (Scotland and northern England, common usage); **hogg-hole** (England, Garner 1984; England [Lakeland], Rollinson [1969] 1972); **lunky/lunkie hole** (England, Garner 1984; Scotland, common usage); **pen hole** (Scotland, Hart 1980); **run hole** (Scotland, O Borchgrevink. 1982); **sheep creep** (Ireland, MacWeeny and Conniff 1986); **sheep gap** (Ireland [Co. Clare], O Danachair 1957); **smoot/smoose/smout** (England, Garner 1984; Scotland and northern England, Hart 1980); **thawl or thirl hole** (Scotland and northern England, Hart 1980).

APRON ROCK: *See* foundation.

ASHLAR: Masonry of squared stones with flat faces laid in horizontal courses with vertical joints.

BASE: *See* foundation.

BATTER: The slight taper of the fence inward toward the top, creating steeply sloped sides (Kentucky, common usage; Scotland and northern England, Hart 1980; Scotland, Callander [1982] 1986; United Kingdom, Brooks [1977] 1989). Synonyms: **camber** (Scotland and northern England, Simkins 1989a); **taper** (Kentucky [Bourbon Co.], Miles 1989; Kentucky [Mercer Co.] Thomas in Mastick 1976; Ohio, Hamilton County Agricultural Society 1830).

BED: The plane of stratification in sedimentary rock; in masonry, a course prepared for a particular, often very large, rock (Kentucky, common usage; United Kingdom, common usage).

BEDDING: *See* foundation.

BEDDING PLANE: The surface of a rock layer in the earth; *see also* grain.

BIND STONE/BINDER: *See* tie-rock.

BLOCK: A large rock or boulder in a fence face (Kentucky, common usage; Scotland, Stephens 1877). Snonyms: **blonk** (Scotland, Brooks [1977] 1989).

BONDER: Rocks extending part way through a fence, serving the same purpose as *tie-rocks; used in regions where large tie-rocks are not available (England [Mendips], Brooks [1977] 1989). Synonym: **keystone** (England [Mendips], Brooks [1977] 1989.

BOUNDARY FENCE: Fence on the line between separately-owned properties. Synonyms: **line fence** (Kentucky, common usage); **march dyke** (Scotland, common usage).

BROKEN COURSING: The bottoms of the rocks do not form continuous horizontal lines, so that the courses are "broken"; pattern contrasts with *straight-coursing. Synonyms: **broken ashlar** (using tooled stone; Kentucky [Franklin Co.], Guy 1989); **random coursing** (Kentucky [Bourbon Co.], T. Soper 1988); **random rubble coursing** (using unworked rock; Scotland, common usage).

BROKEN JOINT: *See* covered joint.

BUCK-AND-DOE: *See* crenellated.

BULLET: A small rock, gathered from fields, used to fill the fence cavity; *see also* spalls (Ireland [Ulster], Evans 1956).

CAMS: *See* coping.

CANT: Unlevel, leaning, sloping, or tilting (New York [regarding Kentucky], *American Agriculturist* 1859).

CAP/CAP STONES/CAPPING: *See* coping.

CAP COURSE: The uppermost horizontal course of the fence; the course on which the *coping rests; *see also* projecting cap course (Kentucky nineteenth and early twentieth century, common usage). Synonyms: **capping base** (Kentucky [Clark Co.], Venable 1988); **coverband** (Scotland and northern England, common usage); **covers** (England, Stephens 1877).

CAP COURSE, PROJECTING: A *cap course extending two to three inches from the face of the fence, having a protruding edge and forming a cornice upon which a *coping usually rests. Synonyms: **bell** (Kentucky [Franklin Co.], Guy 1989); **flat cap** (Kentucky [Scott Co.], Harp 1988); **lip cap** (Kentucky [Scott Co.], Waugh 1988); **overhang** (Kentucky [Franklin Co.],

Guy 1989); **overlap belt** (Kentucky [Fayette Co.], Hayden 1989); **set-out course** (Kentucky [Fayette Co.], Hockensmith 1989); **shelf cap** (Kentucky [Fayette Co.], Moreland 1989).

CASTELLATED: *See* crenellated.

CHAIN: A traditional unit of measurement equaling twenty-two yards.

CHEVALDEFRISE: *See* crenellated.

CHEVAUX DE FRISE: A thirty-foot-wide band of upright stone blocks with sharp ridges on the tops, a wall of defense at ancient forts against attackers on horseback or on foot; compare with *crenellated (Ireland [Inishmore], Richard Conniff in MacWeeny and Conniff 1986).

CHINKING: Small rocks in the outside fence face, wedged between the larger coursed rocks to hold them in place; compare with *filling; *see also* spalls (Kentucky, common usage). Synonyms: **chinking up or pointing up** (Kentucky [Bourbon Co.] Miles 1989); **front pinning** (Scotland, common usage); **pinners** (England [Cotswolds], Brooks [1977] 1989); **pinning/pins** (Ohio, *American Agriculturist* 1830; United Kingdom, common usage); **wedge** (Kentucky [Woodford Co.], Brown 1990; England [Cotswolds], Brooks [1977] 1989; northern England and Scotland, Hart 1980).

CHIPPINGS/CHIPS: *See spalls.*

CLEARANCE WALL: Fence built of rock gathered from the fields. Synonyms: **accretion wall** (northern England, Brooks 1989); **clearance wall** (Scotland and England, common usage); **consumption dyke** (Scotland, common usage).

COCK AND HEN: *See* crenellated.

COMB/COMBERS: *See* coping.

COPE/COPE STONES: *See* coping and cap course.

COPING: The fence top, usually formed by placing rocks on edge, either vertically or at an angle (Kentucky, common usage; northern England, common usage). Sometimes, more recently, the coping is referred to as the "cap" or "capping." Some copings consist of very large solid blocks laid horizontally on top of the fence. Synonyms: **cam/cam stone** stone (northern England, common usage); **cap** (Kentucky, occasional modern usage; Scotland, Hamilton Co. [Ohio] Agricultural Society 1830); **cap stones** (Indiana, Mastick 1976); **capping** (Kentucky, occasional modern usage; Kentucky [Woodford Co.], *Valley Farmer* 1857; England, Manners 1974; Scotland and northern England, Simkins 1989a); **comb** (Kentucky [Mercer Co.], W.B. Smith 1989; Kentucky [Anderson Co.], Worford 1988; England [Cotswolds and the southwest], Hart 1980); **combers** (England [Cotswolds], Rainsford-Hanney 1957); **cope** (England, Stephens 1877; Ohio, Hamilton Co. Agricultural Society 1830; Scotland and northern England, common usage); **cope stones** (Kentucky [Bourbon Co.], Miller 1989; Scotland and northern England, common usage); **crown** (northern England, Hart 1980); **soldier cap** (Kentucky, Brown 1989); **soldier course** (Kentucky [Anderson Co.], Cutsinger 1989; Kentucky [Fayette Co.], Hockensmith 1989); **soldier row** (Kentucky [Jessamine Co.] Hayden

1988); **tie-rocks** (Kentucky [Woodford Co.], Gormley 1988); **top** (Scotland, O Borchegrevink 1982); **topping** (England and Scotland, Hart 1980); **top-stones** (England [Yorkshire], Brooks 1977).

COURSE: Parallel rows of rocks, usually laid horizontally. Synonym: **tier** (New York [re. Kentucky], *American Agriculturist* 1859).

COURSED ASHLAR: Squared stones in *straight coursing.

COVER/COVERBAND: *See* cap course, projecting.

CRAZY WORK: *See* patterns of mortared work.

CREEK ROCK/CREEK STONE: Rocks gathered from creek *beds,* having rounded edges and smoothly worn surfaces; different from *quarried rock from a creek *bank*. Synonym: **water-worn** (Kentucky, Scotland, and northern England, common usage).

CREEL: To lean, tilt, or collapse; to fall into a state of disarray (Kentucky [1989], E. Conner 1989; Kentucky [Bourbon Co.], Miles 1989; Scotland, Webster's New International Dictionary 1923). Synonyms: **fall out** (Kentucky [Bourbon Co.], T. Soper 1988); **rack** (Kentucky [Mercer Co.], Higgins 1988); **slump** (England [Cornwall], Brooks [1977] 1989).

CREEP-HOLE: *See* animal gap.

CRENELLATED: Coping pattern having horizontal and vertical stones alternating, forming a short-tall-short pattern; popular around the turn of the century in Kentucky but used earlier in Kentucky, Ireland, and Britain, and still used occasionally in Britain; *see also* chevaux de frise. Synonyms: **buck-and-doe** (England [Yorkshire], Brooks 1977; Scotland, Callander [1982] 1986); **castellated** (Kentucky, common usage); **cock-and-hen** (England [Cotswalds], Hart 1980); **chevaldefrise** (New York [re. Kentucky], *American Agriculturist* 1859); **one up, one down** (England, Hartley 1951).

CRIPPLE: *See* animal gap.

CROSS STONE: *See* tie-rock.

CROSSED JOINTS: *See* joints, crossed.

CROWN/CROWN STONES: *See* coping.

CUNDIE: *See* water gap.

CULVERT: *See* animal gap and water gap.

CUT STONE: Quarried stone that has straight edges shaped with a power saw; may have a rough or "rusticated" face.

DIAGONAL COURSING: Fences laid with rock edges or most narrow faces towards the ground and coursed at an upright angle between 25 and 40 degrees from vertical. Synonyms: **edge fence** (Kentucky [Anderson Co.], Worford 1988); **rick-rack** (Kentucky [Scott Co.], Waugh 1988); **rip rap** (Kentucky [Mercer Co.], W.B. Smith 1989; Kentucky [Clark Co.] Venable 1989); **rough and ready** (Kentucky [Bourbon Co.] Letton 1989); **stacked fence** (Kentucky [Franklin Co.] Guy 1989); **stand up fence** (Kentucky [Scott Co.], Harp 1988); **up-and-down fence** fence (England [Cornwall] 1989, Brooks [1977] 1989); **vertical coursework** (Irish Sea [Isle of Man] Rainsford-Hannay 1957).

DITCH: In Ireland, a bank, wall, fence, or other raised barrier.

DOUBLE-FACED FENCE: The most common rock fence type in Kentucky, having a coursed face on both sides, which are connected to each other by tie-rocks, the cap course, and the coping; *see also* rock fence. Synonyms: **double ditch** (Ireland [Ulster], Evans 1956); **double/double dyke** (Scotland and England, common usage).

DOUBLING STONE: *See* tie-rock.

DRAIN DAM: An embankment to prevent erosion (Kentucky, common usage). Synonyms: **catch dam** (Kentucky [Clark Co.], Venable 1989); **fleet dyke** (Scotland, Callander [1982] 1986); **retaining dyke/wall** (Great Britain, common usage).

DRESSED STONE: Quarried rocks hand-tooled into precise shapes, usually with squared corners and edges on the exposed *face, used for *ashlar masonry.

DRY JOINTS/DRY SEAMS: Fence joints having no mortar; the joints are "open," with rocks resting directly on one another.

DRY-LAID FENCE: Rock fence having no mortar (Kentucky, common usage; northern England [17th c.], Raistrick [1946] 1988). Synonyms: **dry fence** (Ireland [Ulster], Evans 1956); **dry stone dyke** (Scotland, Stephens 1877; Scotland [ancient term], Tufnell 1990); **dry stone hedge** (England [Cornwall], Hart 1980); **dry stone wall** (Ohio, Hamilton Co. Agricultural Society 1830; England, common usage); **drystane dyke** (Scotland, Hart 1980); **rip rap** (New England, Meredith 1951).

DRY-ROCK FENCE: *See* dry-laid fence.

DYKE/DIKE: *See* rock fence.

EDGE FENCE: *See* diagonal coursing.

FACE: (noun) One of the two exposed, outer sides of a rock fence; also the exposed working wall in a quarry; (verb) to apply rock or stone veneer to a concrete block core or wall (Kentucky, common usage [all meanings]; Scotland and England, common usage).

FALL-OUT: *See* gap.

FEATHERS AND PLUGS: Half-circular steel shims and pointed wedges driven into drilled holes to break off blocks of rock (Kentucky, common usage). Synonyms: **feathers and wedges** (United Kingdom, Brooks [1977] 1989); **plug and feathers** (Scotland, common usage).

FENCE: A structure used for an enclosure, barrier, or boundary; loosely used to include walls, hedges, banks, ditches, and dykes; in Great Britain, now usually a wire structure.

FENCE MASON: A stonemason who specializes in rock fence construction. Synonyms: **dyker** (Scotland, common usage); **waller** (England, common usage).

FIELD ROCK/FIELDSTONE: Rocks gathered from the fields after clearing or plowing (field rock: Kentucky, common usage; fieldstone: New England, common usage). Synonyms: **cloddings/field cloddings** (Scotland, common usage for poor quality fieldstone); **pick-up rock** (Kentucky [Mercer Co.], Kelly 1989a; Kentucky [Clark Co.], Venable 1989).

FILLER/FILLING: Spalls packed in the fence cavity; compare with *chinking (filling: northern England, Brooks [1977] 1989; Kentucky [Fayette Co.], Brown 1989; Scotland and northern England, Hart 1980). Synonyms: **fill** (England, Garner 1984); **fill-in/filled in** (Kentucky [Bourbon Co.], Miles 1989); **hearting** (England, Stephens 1877; Scotland, Hamilton Co. [Ohio] Agricultural Society 1830; Scotland and England, common usage); **infill** (England [Mendips], Brooks [1977] 1989); **packing** (Scotland, O Borchegrevink 1982).

FLAG PATTERN: *See* patterns of mortared work.

FLAT CAP: *See* cap course, projecting.

FLAT FENCE: *See* straight coursing.

FOOTER/FOOTING: A poured concrete base upon which a *foundation is built.

FOUNDATION: The below-ground base upon which a fence is constructed; in modern usage, the foundation includes the footing and lower courses (Ohio, Hamilton Co. Agricultural Society 1830; England [Yorkshire], Brooks 1977; Kentucky [Woodford Co.], *Valley Farmer* 1857). Synonyms: **apron** (Kentucky [Bourbon Co.] T. Soper 1988); **base** (Kentucky, common usage); **bedding** (Kentucky [Mercer Co.], W.B. Smith 1989); **footer** (Kentucky [Franklin Co.], Guy 1989; Kentucky [Bourbon Co.], Miles 1989); **footing** (Scotland and northern England, Hart 1980; England, Garner 1984); **found/foond** (Scotland, Hart 1980; Scotland 1986, Callander [1982] 1986); **foundation stones** (Kentucky [Bourbon Co.], Miller 1989).

FREESTONE: Stone that can be split in any direction.

GAP: An unintentional break in a fence because of a defect or damage, *see also* creel; or a purposeful opening in a fence, *see also* animal gap and water gap (Scotland and northern England, Hart 1980; northern England, Brooks 1989). Synonyms: **breach** (Scotland and northern England, Hart 1980); **slap** (Scotland, Hart 1980).

GRAIN: Particles of crystals in rock, lying parallel to the *bedding plane; also, the direction a rock will split.

HA-HA: A wall below a terrace or embankment, originally to confine stock without impeding a view; *see also* retaining wall. In Britain, a ha-ha is fronted with a depressed ditch.

HEAD: The end of a fence section; two adjoining fence heads butted together without forming a break sometimes mark a change of ownership (England, Garner 1984; England [Lakeland district] Rollinson [1969] 1972; Kentucky [Bourbon Co.], Miller 1989; Scotland and northern England, Hart 1980). Synonyms: **cheek** (Scotland and northern England, Hart 1980); **sconcheon/scuncheon** (Scotland, Hart 1980).

HEADER: *See* tie-rock.

HEARTING: *See* filling.

HOGG-HOLE: *See* amimal gap.

INFILL: *See* filling.

JOINT: The vertical seam between rocks (Kentucky, common usage; Scotland and northern England, Hart 1980).

JOINT, ALIGNED: A place where the *joint of one course is directly above another, creating a vertical seam spanning two or more courses. Synonyms: **risband** (Great Britain [Upper Annandale], Hart 1980); **run** (Vermont, Vivian 1976); **running joint** (Kentucky, common usage); **stacked joint** (Kentucky [Franklin Co.], Guy 1989); **straight joint** (Scotland and northern England, Hart 1980).

JOINT, COVERED: The *joint between two rocks spanned by a single rock in the course above; the opposite of an *aligned joint (Kentucky, common usage). Synonyms: **break joint** (Kentucky [Franklin Co.], Guy 1989); **breaking joints** (England, Garner 1984; United States, Martin [1887] 1974; Scotland, Callander [1982] 1986); **broken joint** (Indiana, Mastick 1976); **crossed joints** (England [Yorkshire], Brooks 1977); **two-over-one joint** (New Hampshire, Fields 1971).

JUMPER: A tall face stone, often at a post corner; may span two courses; may be laid against the grain; *see also* shiner (Kentucky [Franklin Co.], Guy 1989; England [Cotswolds], Brooks [1977] 1989).

KEYSTONE: The locking stone at the apex of an arch; *see also* bonder.

LEDGE QUARRY: Quarry opened on a hillside where soil erosion exposes underlying rock strata.

LEDGE ROCK: Rock removed from ledges or outcroppings in hillsides or creek banks; *see also* quarried rock.

LIFT: the point from which the fence is built upon the foundation (Scotland, common usage).

LINTEL: A rock or stone slab bridging an opening and supporting the courses above.

LIP CAP: *See* cap course, projecting.

LUNKY HOLE: *See* animal gap and water drain.

MORTARED FENCE: Fence having joints built or sealed with cement mortar. Synonym: **wet wall** (Kentucky, R.R. 1956).

NICKERS: Wedge-shaped stones used to bring the course up to level; compare with *chinking (Scotland, Brooks [1977] 1989).

ONE UP, ONE DOWN: *See* crenellated.

OPEN JOINT/OPEN SEAMS: *See* dry joints.

OUTCROPPING: A place in a hillside where topsoil erosion exposes underlying rock ledges; also called a ledge.

OVERLAP BELT: *See* cap course, projecting.

PACKING: *See* filling.

PATTERNS OF MORTARED WORK: *Crazy work:* irregular, unshaped rocks laid on edge, against the grain, with no coursing pattern; *flag pattern:* in mortared work, irregular or uncoursed work laid like flagstone; *repeating:* pattern having stone sizes and a coursing pattern occurring at regular intervals; *two-to-one:* a repeating pattern in which large stones span two or three courses at regular intervals and in regularly alternating rows, built of square-edged stone having a rough face.

PERCH: A unit of measurement of twenty-four and three-fourths (or twenty-five) cubic feet; a volume one foot tall and one foot deep by twenty-five feet long; sometimes one foot high by eighteen inches deep by sixteen feet long.

PIER: *See* pillar.

PICK-UP ROCK: *See* field rock.

PILLAR: A stone post at the end of a fence section or at each side of a water gap or gateway; also called post, pier, or column (Kentucky [Madison Co.] Rusticus 1859). Synonyms: **abutment** (Kentucky [Bourbon Co.], Clay 1849); **column post, entrance post, main gate post, pillar** (Kentucky [Scott Co.], B. Waugh 1989).

POINTING: The final mortaring of joints (Kentucky, common usage; United Kingdom, common usage).

POINTING, RAISED: On mortared fences, mortar joints finished in square or angled projecting shapes. Patterns of raised pointing: *fork point* (Kentucky [Mercer Co.], Kelly 1982; *raise point* (Kentucky [Franklin Co.], Guy 1989); *ribbon pointing* (Kentucky, common usage).

POLE: A unit of length measuring sixteen and one-half feet; similar to a *rod.

POST: *See* pillar.

QUARRIED ROCK: Rock excavated from an opening in a hillside; *see also* ledge rock.

QUARTER: English unit of height equaling nine inches (one quarter of a yard).

RACK: *See* creel.

RANDOM ASHLAR: A coursing pattern using squared stones not laid in continuous courses.

RETAINING WALL: Coursed rock-work built to support an earthen bank or terrace (Kentucky, common usage; Scotland and northern England, Hart 1980). Synonyms: **ha-ha** (Kentucky, common usage); **single faced dyke/wall** (Scotland, Stephens 1877); **sunk dyke** (Scotland, Callander [1982] 1986; more often refers to a dyke/wall suffering subsidence, as in a bog [Tufnell 1990]).

RIBBON POINTING: *See* pointing, raised.

RICK-RACK: *See* diagonal coursing.

RIP RAP: *See* diagonal coursing, and dry-laid.

RISBAND: *See* joint, aligned.

RIVER ROCK: Bluegrass term for Ordovician period Tyrone limestone.

ROCK FENCE: A fence built of rocks which have been gathered from fields or creek bottoms, or of undressed ledge or quarried rock; also, the Kentucky term for *stone fence and *stone wall (Kentucky, modern common usage for rock or stone fence). Synonyms: **ditch** (Ireland and parts of Wales, Hart 1980); **dyke** (Scotland and northern England, common usage); **stone fence** (Kentucky, mid-nineteenth century usage for rock or stone fence); **stone wall** (England, common usage).

ROD: Unit of measure, 5.5 yards, 16.5 feet, or 5.029 meters.

ROOD: A British unit equal to one-fourth acre, or a British unit equal to seven or eight yards, or sometimes the same as *rod. Other uses of the term: rood = thirty-six yards in length, forty-five inches tall (England, Stephens 1877); rood = measure of length, six yards in granite districts, seven yards in limestone districts, generally a day's work for one man (Scotland [Galloway], Rainsford-Hannay 1957).

ROUGH AND READY: *See* diagonal coursing.

RUBBLE: Rough, untrimmed rock; *see also* field rock and spalls.

RUN HOLE: *See* animal gap.

RUNNING JOINT: *See* joint, aligned.

RUSTICATED: *See* cut stone.

SCARCEMENT: The inset between the outer edge of the foundation and the bottom of the fence, usually two inches (Scotland, common usage).

SCRAP ROCK: *See* spalls.

SEAMS: The spaces between the courses and joints of the rocks.

SET OUT CAP/SET OUT COURSE: *See* cap course, projecting.

SHEEP CREEP/SHEEP GAP: *See* animal gap.

SHELF CAP: *See* cap course, projecting.

SHINER: A rock placed with its grain vertical in an otherwise flat-laid course; *see also* jumper (Kentucky [Fayette Co.], Giles 1988). Synonym: **jumper** (Kentucky [Mercer Co.], Higgins 1988; England [Lake District], Bodman 1984).

SHIVERS: *See* spalls.

SINGLE FACED WALL: *See* retaining wall.

SKINTLED WORK: *See* tie-rocks, projecting.

SLUMP: *See* creel.

SMOOT: *See* animal gap.

SOLDIER COURSE/SOLDIER ROW/SOLDIERS: *See* coping.

SPALLS: Small pieces of rock or chips that are debris from *dressed rock, used for fill on the inside of the fence (then called *filling or hearting), for *chinking, and for fill in a fabricated coping (Kentucky, common usage). Synonyms: **chippings** (England, Manners 1974); **chips** (published in New York, Martin [1887] 1974; Kentucky [Mercer Co.], W.B. Smith 1989); **rubble** (Kentucky, modern usage); **scrap/scrap rock** (Kentucky, common usage); **shivers** (Scotland, Stephens 1877); **spralls** (Kentucky [Franklin Co.], Guy 1989); **waste** (Kentucky [Woodford Co.], Brown 1990).

SPRALLS: *See* spalls.

STACKED FENCE: *See* diagonal coursing.

STILE: Large rocks or stones protruding from the fence face, forming steps (Kentucky, common usage; Scotland and northern England, Hart 1980). Synonyms: **climbing stones** (Kentucky [Woodford Co.], Higgs 1989); **steps** (Kentucky, common usage).

STONE FENCE: A fence built of rocks that have been quarried and *dressed. In other regions of the United States "stone fence" or "stone wall" may be the term for what in Kentucky is known as a "*rock fence."

STONE SLED: A flat-bedded sled on wooden runners, usually reinforced with metal rims, used for hauling rock from fields or quarries to the fence site. Synonym: **sledge** (Scotland, Stephens 1877).

STONE WALL: *See* rock fence.

STOOP/STOUP: An upright monolith set into the ground against the wall head at a gate or open stile (Scotland and northern England, Hart 1980; Scotland, Callander [1982] 1986).

STRAIGHT COURSING: All rocks in each course are the same height, forming continuous horizontal lines, although the courses may be of varying heights. In regions where diagonal coursing is common, straight coursing is referred to as "flat fence."

STRAIGHT JOINT: *See* joint, aligned.

THAWL OR THIRL HOLE: *See* animal gap.

THROUGH BAND/THROUGH COURSE: *See* tie course.

THROUGH ROCK/THROUGH STONE/THROUGHS: *See* tie- rock.

TIE COURSE: A course composed entirely of *tie-rocks. Usually the tie course is the top horizontal course, in which case it is the same as the *cap course (Kentucky, most common usage). Synonym: **through band** (Scotland and northern England, Hart 1980; England, Stephens 1877; Scotland, O Borchegrevink 1982).

TIE-ROCK: A rock laid with its longest dimension running entirely through the fence as a header, having the effect of tying the two faces of the fence together. *Coping-rocks and *cap courses that cover the full width of the fence also act as tie-rocks. A course composed entirely of tie-rocks is a *tie course (Kentucky, most common local usage). Synonyms: **bind stone** (United States [pub. in New York], Martin [1887] 1974); **binder** (central Kentucky, *American Agriculturist* 1859; Kentucky [Woodford Co.], *Valley Farmer* 1857; Scotland, Hamilton Co. [Ohio] Agricultural Society 1830); **bonder** (England [Cotswolds and Mendips], Brooks [1977] 1989); **doubling stone** (Scotland, Brooks [1977] 1989); **locking stone** (United States, Burns 1988); **through stone/throughs** (England, common usage; Kentucky [Bourbon Co.], Miller 1989); **tie** (Scotland, Hart 1980; Scotland, Callander [1982] 1986); **tie-in rock** (Kentucky [Anderson Co.], Cutsinger 1989).

TIE-ROCKS, PROJECTING: *Tie-rocks that protrude from the fence face a few inches because their ends have not been squared or broken off. Synonyms: **extended ties** (Kentucky [Clark Co.], Venable 1988); **skintled work** (Kentucky [Fayette Co.], Hockensmith 1989).

TIE-IN ROCK: *See* tie-rock.

TIE-STONE: *See* tie-rock.

TIER: *See* course.

TOP/TOPSTONES: *See* coping.

TRACE-WALLING: Placing the longest length of the stones along the face of the wall instead of into it (northern Britain, Simkins 1989c).

TWO-OVER-ONE JOINT: *See* joint, covered.

TWO-TO-ONE PATTERN: *See* pattern of mortared work.

UP-AND-DOWN FENCE: *See* diagonal coursing.

VENEER: A rock or stone facing applied to a concrete or concrete block core.

WATER DIVERTER: Purpose-built ridge in a road to direct rainwater to side ditches or drains. Synonym: **dead Irishman** (Kentucky [Scott Co.], Goodman 1989).

WATER DRAIN: A small opening in a fence to allow water to run through. Synonyms: **conduit** (Kentucky, Soper 1990; Scotland, Cairns [1975] 1986); **cundie** (Scotland and northern England, Hart 1980; Scotland, Callander [1982] 1986); **lunky hole** (Scotland, O Borchgrevink 1982; Scotland [Galloway], Rainsford-Hannay 1957); **water smoot** (Scotland and northern England, Hart 1980); **waterpen** (Scotland and northern England, Simkins 1989a); **weep hole** (Kentucky [Bourbon Co.], E. Conner 1989).

WATER GAP: An opening built in a fence at a watercourse; the opening is the width of the water-course; wooden gates are hung between pillars at each side of the opening.

WATER-WORN ROCK/STONE: *See* creek rock.

WEDGE: *See* filling.

WEEP HOLE: *See* water drain.

Notes

1. Landforms, Rock, and Quarrying

1. "The Bluegrass" is a curious name for a place. It is not known whether the grass from which central Kentucky took its name was native or imported (Campbell 1980). Reports by early travelers do not solve this mystery. James Nourse, on his trip to Kentucky in 1775, mentioned in his journal the "blue grass" he found ([1775] 1925, 252), but this is not the clear identification that it seems since at that time there were two dozen different grasses known as blue grass in America and Europe (Carrier and Bort 1916, 257). Almost sixty years later Kentuckian Henry Clay, statesman, farmer, and amateur scientist, said that "our blue grass is very different from the grass which is known by that name in Maryland and Virginia. It is the green sward or speargrass of England. When Kentucky was first settled small patches of it were found . . . as the Country opened and improved, its value became known, and it was extended and diffused" (Dunbar 1977, 522). As the Bluegrass region became known for the predominance of this favored pasture grass, the popular name for the grass and the land where it flourished became synonymous.

2. The *American Agriculturist* reported that woodland pasture was created "by merely underbrushing and thinning out a few of the trees of the original forest. These are usually termed wood-land pastures in Kentucky; yet in most instances, they better deserve the name of park, than many of those on noblemen's estates in Europe" (1843, 323).

3. Because many sources in this essay are interviewees and correspondents, full information on them is included in the Reference List, in order that parenthetical references may be brief.

4. John Henry, hero of the folk ballad, was a steel-drivin' man who, according to tradition, wielded a thirty-pound hammer with a nine-foot handle.

2. Rock Fence Construction

1. Builders of rock fences in Kentucky are known as stonemasons, fence masons, or rock fencers. This terminology becomes confusing when the person building a *rock* fence is referred to as a *stone*mason. The term "stonemason" also does not differentiate between a "dry" mason, who builds dry-laid fence, and a "wet" mason, who uses mortar, although the crafts are quite different. The term "apprentice" is also imprecise. Although county order books contain records of formal apprenticeship to stonemasons engaged in construction of buildings, no known record identifies apprenticeship to a fence mason. The term therefore may be casually employed in the fence-building trade.

2. During the seventeenth and eighteenth centuries, the term "planter" referred to a person who inhabited (i.e., was planted in) British colonial lands such as Ireland and the American colonies. In America, the term was synonymous with the present meaning of

"farmer" instead of its somewhat later meaning: large landowner and slaveholder (Isaac 1982, 16). In the late eighteenth and early nineteenth centuries in Kentucky, the term "plantation" referred to what is now called a farm. In wills of this period, the homeplace is often referred to as "the plantation on which I reside." This is true of people both of great and modest wealth. Records from this period refer to holdings of as little as two hundred acres as "plantations" (Murray-Wooley 1988). "Farm," however, came to be the popular term by the mid-nineteenth century. Only a few men are listed by occupation in the 1850 census as planters; by that time most agriculturists were classified as farmers.

3. We have used the term "mason" for simplicity to indicate the builder of the fence. The builder could, however, have been the farmer-owner and his laborers, several hired masons, or a mason and his helpers or apprentices.

4. It seems logical to speak of the "thickness" of a rock or rock bed. We have instead used "height" for both the vertical dimension of a single rock and the vertical dimensions of rock beds and fences, to avoid the possible confusion of thickness with width. We have used "width" to indicate the cross-sectional dimension of a fence from one face to the other and "length" to indicate the horizontal facade dimension.

5. Sources consulted for this essay cover 1777 to the present. Names of the masons and quarriers from these records are listed in Appendix 1. Although the list contains over twelve hundred names of historic and contemporary stonemasons, it represents only a fraction of the total number of central Kentucky fence masons.

6. Figures of the Buckner, Soper, and Butts farms are diagrammatic, portraying only existing rock fencing and farm boundaries. They do not depict existing barns, springs, watercourses, and other fencing, nor do they indicate demolished rock fences, buildings, or former farm boundaries.

3. Bluegrass Fencing Traditions

1. "Ditch" is used here in the northern Irish sense meaning a raised earthen wall. Hedges planted atop raised earth or sod ditches were common as fencing in Ulster in the eighteenth century (Evans 1956; 1957, 109-10).

4. Origins of Fence Masons

1. While census records indicate that most Irish turnpike workers and stonemasons came to Kentucky after the potato famine, there is evidence that this wave of migration was a continuation of an already established pattern. A letter from an Irishman written before 1818 refers to the great number of Irishmen employed on American turnpikes (Melish [1812] 1818, 622). Further, Godfrey Vigne observed in 1832, well before the famine emigration, that in America "the Irish . . . find employment as roadmakers" (85). For more information regarding the Irish exodus to America, see Cecil J. Houston and William J. Smyth, *Irish Emigration and Canadian Settlement: Patterns, Links, and Letters,* and Kerby A. Miller, *Emigrants and Exiles,* chapter 7.

2. Tilford Huff recorded additional information about the Rose Hill neighborhood in "On banks of Chaplin," a series of articles in the *Harrodsburg* [Kentucky] *Herald,* the last one of which appeared about 1909 (J. Huff 1989).

3. Census takers listed the inhabitants of a single household together in the enumeration. They proceeded from house to neighboring house, so that the proximity of names in the census list reflects the order in which the census taker recorded them, but, more important, indicates that they clustered together in ethnic neighborhoods in central Kentucky towns and villages such as Irishtown in Lexington. Sometimes an entire household of stonemasons lived next door to a crew of turnpike laborers.

4. Fence masonry was taught at Tuskegee as late as 1952 (O'Connor 1952, 6).

5. Buildings having masonry by the Guy family include Keeneland clubhouse in Fayette County, the home of Hugh Meriwether in Lexington, the home of George M. Chinn in Mercer County, and the wings of the A.B. Gay house in Woodford County.

6. We did not read the 1860 or subsequent censuses of Boyle County, where the Meaux family lived.

7. For more on Celtic cultural history, see William Rollinson, *A History of Man in the Lake District.*

8. This group includes Ulster-Kentuckians with Irish, Scottish, Welsh, English, and French surnames, reflecting the diverse origins of Northern Irishmen (Murray-Wooley 1988). One modern American writer (Fischer 1989) groups all "backcountry" American colonists as "North Britains" in his thesis that the northern English, Scots, and northern Irish, while of different ethnic stocks, shared a single cultural region in the lands surrounding the north Irish Sea—a view not accepted by northern Irish and Scottish scholars (Blaustein 1990; Cowan 1990)—and that they became the dominant English-speaking culture on the American frontier.

9. The presence or absence of rock fences was noted at 168 sites containing stone houses built between 1780 and 1830 for which the ethnicity of the original owner is known. Rock fences currently survive at 75 of these and are absent at 93. The families of the owners of 92 of these 168 sites had last resided in Ireland (primarily Ulster) before immigration to America, 27 were from England, 21 from Wales, 13 from Scotland, and 16 from France, Germany, Holland, Switzerland, or Poland. The stone house properties originally owned by families from Scotland and Ireland have significantly higher incidence of rock fences than those associated with other ethnic groups:

Scotland—62 percent
Ireland—50 percent
Wales—43 percent
Other European countries—31 percent
England—29 percent

The number of stone houses that once had rock fences is probably much higher than is known since an estimated 90 percent of Kentucky's rock fences no longer stand.

10. Frederick Rainsford-Hannay was founder of the Stewartry of Kirkcudbright (Scotland) Stone Dyking Committee in 1938, forerunner of the present-day Dry Stone Walling Association of Great Britain.

5. *Brutus Clay's Auvergne*

1. The Clay Papers are held at the M.I. King Library Special Collections at the University of Kentucky and are organized into chronologically ordered files. A contract or letter from 1819, for example, is filed with other materials from the same year.

2. In 1854, Mary McAboy, a friend of Brutus Clay's second wife, Ann, suggested the name "Auvergne" for the plantation because of its evergreen setting (McAboy 1859; Whitley 1956, 12).

3. Frances Thornton Spengler of Winchester, Kentucky, related the following: her father was Francis Aloyisus Thornton, although he informally changed his middle name to Andrew. His wife's maiden name was Ahern, and she was of Irish descent, distantly related to his grandmother. The name Francis goes "way back" in their family, and Spengler was named for her father, with the feminine spelling "Frances." Her grandfather, Michael Francis Thornton, worked as a tenant on the McCormick farm in central Kentucky and after that on the Clay farm, saving his money until he was able to purchase a farm in Fayette County on the Versailles Pike. One of his sons, Ed (age 95 in 1990), still

occupies this farm. Her great-grandfather, William Bernard Thornton, came to Kentucky from County Cork, Ireland, during the potato famine with four or five of his brothers. William's wife was also an Ahern. The Thornton brothers separated after they came to America, with some of them going to Bourbon County, Kentucky, near Paris. All were tenant farmers when they arrived and could barely sign their names (see the Clay and Thornton agreement of 1884, 12) (Spengler 1990).

4. These fences were built of shaped rock (stone). Brutus Clay's records sometimes refer to them as stone fences and sometimes as rock fences. We have used the terms interchangeably in this chapter.

6. Change and Legacy

1. In addition to listing men as "rockmasons" or "stonemasons," the census often records related occupations. In Mercer County, a man is listed as a "rock blaster" in 1860, and in 1870, a Scott County black man's occupation is recorded as "works in rock." The Woodford County census of 1900 is the first to specifically identify a person's occupation as "works a rock crusher," although there and in Scott County, as in previous years, the occupations of other men were "rock breakers." In Woodford County in 1910, one man was listed as a "rock crusher manager," and in Mercer County that same year, one man was a "rock crusher engineer" and another a "rock crusher laborer."

Reference List

Akenson, Donald H. 1984. *The Irish in Ontario*. Kingston: McGill and Queen's Univ. Press.

Algood, Alice W. (Tennessee Historic Preservation Office). 1989. Letter to authors, 18 Sept.

Allen, James Lane. 1892. *The Blue-grass region of Kentucky and other Kentucky articles*. New York: Harper.

Allen, James P., and Eugene J. Turner. 1988. *We the people: An atlas of America's ethnic diversity*. New York: Macmillan.

Allen, Turner W. 1954. The turnpike system in Kentucky: A review of state road policy in the nineteenth century. *Filson Club History Quarterly* 28 (July): 239-59.

American Agriculturist. 1842. Kentucky farming. 1:67-70, 138-40.

______. 1843. Sketches of the West, number 2. 2:323-25.

______. 1859. Farm fencing. 18:110-11, 174-75.

Anderson, Jonathan Jay. 1851. Letter to Brutus J. Clay. Clay Family Papers. Special Collections, M.I. King Library, University of Kentucky, Lexington.

Armstrong, J.M., ed. 1878. *Biographical enclyclopaedia of Kentucky of the dead and living men of the nineteenth century*. Cincinnati: J.M. Armstrong.

Basham, Kevin (Anderson Co. landowner). 1988. Interview with author, 2 Apr.

Bealeret, John. 1851a. Bill, 5 Jan. Clay Family Papers. M.I. King Library, Special Collections, University of Kentucky, Lexington.

______. 1851b. Bill, 28 Dec. Clay Family Papers. M.I. King Library, Special Collections, University of Kentucky, Lexington.

Beatty, Adam. 1844. *Essays on practical agriculture, including his prize essays*. Maysville, Ky.: Collins and Brown.

Bedinger, D.P. 1851. Letter to Brutus J. Clay. Clay Family Papers. M.I. King Library, Special Collections, University of Kentucky, Lexington.

Bevins, Ann Bolton. 1984a. History of stone fences. *Georgetown* [*Ky.*] *News and Times* 13 Dec.

______. 1984b. Letter to author, 24 Jan.

______. 1988. Letter to authors, 20 Sept.

______. 1989. Letter to authors, Feb.

Bevins, Ann Bolton, and James R. O'Roarke. 1985. *"That troublesome parish."* Georgetown, Ky.: privately printed.

Black, Douglas F.B., Earle R. Cressman, and William C. MacQuown. 1965. *The Lexington Limestone (Middle Ordovician) of central Kentucky*. U.S.G.S. Bulletin 1224-C. Washington: U.S. Geological Survey.

Blaustein, Richard. 1990. Hegemony, marginality and identity reformulation: Further thoughts regarding a comparative approach to Appalachian studies. Paper presented at 8th Ulster-American Heritage Symposium, 2 Aug. at East Tennessee State Univerity, Johnson City.

Bodman, Janet. 1984. *Lake District stone walls.* Lancaster, England: Dalesman.

Bourbon County Agricultural Society. 1852. Receipt to Jim Shy, 10 Sept. Clay Family Papers. M.I. King Library, Special Collections, University of Kentucky, Lexington.

Bourbon County. 1834-1915. County Clerk's Office. Deed books. Paris, Ky.

Brewer, Mrs. Robert (Woodford Co. farm owner). 1988. Interview by authors, 21 Sept.

Brooks, Alan. J. [1977] 1989. *Dry Stone Walling.* Revised by Elizabeth Agate. Wallingford, England: British Trust for Conservation Volunteers.

Brooks, Margaret. 1977. The craft of a dry stone waller. *Folklife* 15:96-105.

Brooks-Smith, Joan E. 1962. *Master index, Virginia surveys and grants, 1774-1791.* Frankfort: Kentucky Historical Society.

Brown, Joe (Woodford Co. masonry contractor). 1990. Interview with author, 20 Sept.

Brown, John Mason. 1870. *Memoranda of the Preston family.* Frankfort: S.I.M. Major.

Brown, Kenneth (Woodford Co. masonry contractor). 1989. Interview with author, 26 Feb.

Buckingham, James Silk. 1842. *The eastern and western states of America.* 3 vols. London: Fisher.

Burns, Susan J. 1988. Rocks of ages. *Garden Design* 7 (Autumn): 8-10.

Burroughs, John. 1895. *Riverby.* Boston: Houghton Mifflin.

Cairns, Robert. [1975] 1986. *Drystone dyking.* Selkirk, Scotland: Biggar Museum Trust.

Caldwell, Clemens (Boyle Co. farm owner). 1989. Interview, 26 Feb.

Callander, Robin. [1982] 1986. *Drystane dyking in Deeside.* Aberdeen, Scotland: published by the author.

Campbell, Julian J.N. 1980. Present and presettlement forest conditions in the Inner Bluegrass of Kentucky. Master's thesis, University of Kentucky.

Carmickle, Wayne (Mercer Co. stonemason). 1990. Interview with author, 24 Oct.

Carrier, Lyman, and Katharine S. Bort. 1916. The history of Kentucky Bluegrass and White Clover in the United States. *Journal of the American Society of Agronomy 8* (July-Aug.): 256-66.

Casey, Russell (Fayette Co. city planner). 1988. Interview with author, 15 Oct.

Cassidy, Samuel M. (Fayette Co. farm owner). 1977. *Lexington Leader* 17 Feb.

———. 1989. Interview with author.

Caton, J.E. (Fayette Co. property owner). 1988. Interview, 19 July.

Cause, Sue. 1980. Lees, Midway stone masons, are dry fence masters. *Lexington Herald-Leader* 24 Dec.

Channing, Steven A. 1977. *Kentucky: A bicentennial history.* New York: Norton.

Chambers, James M. 1854. Work for the month. *Soil of the South* 4 (Dec.): 1-2.

Cherry, Thomas Crittenden. 1930. Robert Craddock and Peter Tardiveau. *Filson Club History Quarterly* 4 (Apr.): 68-90.

Chinn, George M. (Ky. historian). 1984. Interview with author, Oct.

Cincinnati Daily Gazette. 1857. [untitled tribute to Henry Clay] 4 July.

Clark, Bill (Clark Co. stonemason). 1990. Interview with author, Nov.

Clark, Thomas D. 1968. *Kentucky: Land of contrast.* New York: Harper and Row.

———. 1989. Letter to the authors, 22 Mar.

———. 1990. *Footloose in Jacksonian America: Robert W. Scott and his agrarian world*. Frankfort: Kentucky Historical Society.

Clay, Ann M. 1854. Negroe Book. Clay Family Papers. M.I. King Library, Special Collections, University of Kentucky, Lexington.

Clay, Berle (Bourbon Co. farm owner and descendant of Brutus J. Clay). 1990. Interview with authors, 15 Mar.

Clay, Brutus J. 1828-1846. Memorandum Book. Clay Family Papers. M.I. King Library, Special Collections, University of Kentucky, Lexington.

———. 1839-1853. Day Book. Clay Family Papers. M.I. King Library, Special Collections, University of Kentucky, Lexington.

———. 1841. Receipt to Sam H. Chew, 22 Mar. Clay Family Papers. M.I. King Library, Special Collections, University of Kentucky, Lexington.

———. 1846-1877. Memorandum Book. Clay Family Papers. M.I. King Library, Special Collections, University of Kentucky, Lexington.

———. 1849. Receipt from Francis Thornton, 11 Nov. Clay Family Papers. M.I. King Library, Special Collections, University of Kentucky, Lexington.

———. 1850a. Receipt from George M. Lytle, 23 June. Clay Family Papers. M.I. King Library, Special Collections, University of Kentucky, Lexington.

———. 1850b. Receipt from Thomas Malone, 14 Dec. Clay Family Papers. M.I. King Library, Special Collections, University of Kentucky, Lexington.

———. 1854-1875. Day Book. Clay Family Papers. M.I. King Library, Special Collections, University of Kentucky, Lexington.

Clay, Brutus J., and Francis Thornton. 1844. Agreement, April. Clay Family Papers. M.I. King Library, Special Collections, University of Kentucky, Lexington.

———. 1848. Agreement, 9 Oct. Clay Family Papers. M.I. King Library, Special Collections, University of Kentucky, Lexington.

Clay, Brutus J., and Richard Wyatt. 1829. Lease Agreement, 5 Nov. Clay Family Papers. M.I. King Library, Special Collections, University of Kentucky, Lexington.

Clay Family Papers. 1836-1873. M.I. King Library, Special Collections, University of Kentucky, Lexington.

Clay, Green, and John Yeager. 1819. Lease Agreement, 28 July. Clay Family Papers. M.I. King Library, Special Collections, University of Kentucky, Lexington.

Clay, Sidney P., and Emanuel Wyatt. 1825. Lease Agreement, 15 Dec. Clay Family Papers. M.I. King Library, Special Collections, University of Kentucky, Lexington.

Clepper, Henry. 1973. The birth of the CCC. *American Forests* 79 (Mar.): 8-11.

Cobbett, William. 1828. *A year's residence in the United States of America*. London: published by the author.

Colbert, Mrs. Laurence P. (New York State landowner). 1980. Letter to author, 1 Sept.

Coleman, J. Winston. 1953. Lexington as seen by travelers. *Filson Club History Quarterly* 29:267-81.

Conner, Edward (Bourbon Co. fence mason). 1989. Phone interview with author, 26 Feb.

Conner, Nancy J. Underwood (Fayette Co. descendant of mason John Staley). 1989. Letter to authors, 12 Feb.

Cowan, Edward. 1990. Prophecy and prophylaxis: A paradigm for the Scotch-Irish? Paper presented at 8th Ulster-American Heritage Symposium, 2 Aug., East Tennessee State University, Johnson City.

Creech, John (Franklin Co. stonemason). 1990. Interview with author, 18 Oct.
Creger, Elsie Wilson Littral (Clark Co. Wilson family descendant). 1989. Interview with author, 20 Feb.
Crump, M.H. 1895. *Kentucky highways: History of the old and new systems.* U.S. Department of Agriculture, Office of Road Inquiry, Bulletin no. 13. Washington: GPO.
Cumberland [Ky.] Tri-City News. 1971. Cumberland Park helps restore skill of stonemasonry, a waning craft. 24 Oct.
Cummins, John S. 1860. Letter to Richard Heneker, quoted in Cecil J. Houston and William J. Smyth, *Irish Emigration and Canadian Settlement.* Toronto: Univ. of Toronto Press, 1990.
Cuse, John. 1855. Letter to Brutus J. Clay, 7 June. Clay Family Papers. M.I. King Library, Special Collections, University of Kentucky, Lexington.
Cutsinger, Jerry (Anderson Co. stone and fence mason). 1989. Interview with author, 26 Feb.
Danhof, Clarence H. 1944. The fencing problem in the eighteen-fifties. *Agricultural History* 18 (Oct.): 168-86.
Davis, Darrell H. 1927. *Geography of the Bluegrass region.* Frankfort: Kentucky Geological Survey.
Davis, Martin. 1825a. Letter to Green Clay, 24 Jan. Clay Family Papers. M.I. King Library, Special Collections, University of Kentucky, Lexington.
______. 1825b. Letter to Green Clay, 14 Feb. Clay Family Papers. M.I. King Library, Special Collections, University of Kentucky, Lexington.
Dawson, Mr. (Bourbon Co. farmer). 1988. Interview with author, Oct.15.
DeBreffny, Brian. 1982. *Irish family names: Arms, origins, and locations.* New York: Norton.
Draper, Lyman C. [1800s]. Manuscript collection. State Historical Society of Wisconsin, Madison. Microfilm.
Dry Stone Walling Association of Great Britain. 1988. *Building and repairing dry stone walls.* Kenilworth, England: Dry Stone Walling Association of Great Britain.
Dudley, J. 1843. Letter to Brutus J. Clay, 29 May. Clay Family Papers. M.I King Library, Special Collections, University of Kentucky, Lexington.
Duke, Basil W. 1911. *Reminiscences of General Basil W. Duke, C.S.A..* Garden City, N.Y.: Doubleday, Page.
Dunbar, J.S. 1977. Henry Clay on Kentucky Bluegrass, 1838. *Agricultural History* 51 (3): 520-23.
Eads, Sally Sandusky. 1953. New walls from old. *Courier-Journal and Times Magazine* 11 Oct.:34-37.
Estes, Worth. 1958. Henry Clay as a livestock breeder. *Filson Club History Quarterly* 32:351-52.
Evans, E. Estyn. 1956. Fields, fences, and gates. *Ulster Folklife* 2:14-18.
______. 1957. *Irish folk ways.* London and Boston: Routledge and Kegan Paul.
Everman, H.E. 1977. *History of Bourbon County, 1785-1865.* Paris, Ky.: Bourbon Press.
Fayette County. 1846-1850. County Clerk's Office. Order Book. Fayette Co., Ky.
Fidler, Jesse, estate. 1878. County Clerk's Office. Processions Book B:129. Anderson Co., Ky.

Field, E.H. 1837. Letter to the Clay family, 30 Sept. Clay Family Papers. M.I. King Library, Special Collections, University of Kentucky, Lexington.

Fields, Curtis P. 1971. *The forgotten art of building a stone wall: An illustrated guide to dry wall construction*. Dublin, New Hampshire: Yankee.

Fischer, David H. 1989. *Albion's seed: Four British folkways in America*. New York: Oxford Univ. Press.

Fletcher, Austin B. 1907. *The construction of Macadam roads*. U.S. Department of Agriculture, Office of Public Roads, Bulletin 29. Washington: GPO.

Fox, Lester. 1958. Need more acres? Take out stone walls. *Farm Journal* 82 (Dec.): 50-52.

Fox, Lester. 1959. Stone wall removal pays. *Soil Conservation* 24 (May): 225-27.

Franklin [Ky.] Farmer. 1838. For the Franklin farmer. 2 (13 Oct.): 52-53.

Fusonie, Alan. 1977. The agricultural literature of the gentleman farmer in the colonies. In *Agricultural literature: Proud heritage—future promise*, edited by Alan Fusonie and Leila Moran, 33-55. Washington: Association of the National Agricultural Library and the United States Department of Agriculture.

G. 1836. Stone walls. *Farmer's Cabinet*. 1 (1 Sept.): 57.

Garner, Lawrence. 1984. *Dry stone walls*. Aylesbury, England: Shire Publications.

Garrard, Charles T. [1853] 1931. Diary. *Register of the Kentucky Historical Society* 29:400-15.

Giles, Ben (Fayette Co. farm manager). 1988. Interview with author.

Glasscock, Ninie. 1990. Fence-builder at Simmstown Manor is fifth generation. *Springfield [Ky.] Sun* 25 July.

Goodman, Dan (Clark Co., descendant of fence mason Thomas Barrett). 1989. Interview, 7 Mar.

Gormley, Paul J. (Fayette Co., descendant of fence masons Michael and James Gormley). 1987. Interview with author, 12 July.

———. 1988. Interview with authors, 25 Sept.

———. 1990. Interview with author, 17 Sept.

Gregory, Howard (Mercer Co. descendant of fence mason John Franklin Asher). 1990. Letter to authors, 22 Aug.

Guy, John H. III (Franklin Co. fence masonry contractor). 1989. Interview with authors, 3 May.

———. 1990. Interview with author, 14 Oct.

Hall, Eli (Coldstream Farm, Carnahan House resident manager). 1989. Letter to author, 4 Sept.

Hamilton County [Ohio] Agricultural Society. 1830. Fences. *Western agriculturist and practical farmer's guide*. Cincinnati: Robinson and Fairbank.

Hardin, Bayless E. 1981. The Brown family of Liberty Hall. *Genealogies of Kentucky families*. Baltimore: Genealogical Publishing.

Harp, Paul (Fayette Co. masonry contractor). 1985. Interview with author, 25 Oct.

———. 1988. Interview with author, 11 Oct.

———. 1990. Interview with author, 2 Oct.

Harrison, Lowell H. 1970. Kentucky's agriculture, 1879: A British report. *Filson Club History Quarterly* 44:273-81.

Hart, Edward. 1980. *The dry stone wall handbook*. Northhampton, England: Thorsons.

Hartley, Dorothy. 1951. *Made in England*. London: Methuen.

Hayden, Mexico (Jessamine Co. stonemason). 1988. Interview with author, 11 Oct.

Henlein, Paul C. 1959. *Cattle kingdom in the Ohio Valley, 1783-1860*. Lexington: Univ. of Kentucky Press.
Herriot, James, and Derry Brabbs. 1978. *James Herriot's Yorkshire*. New York: St. Martin's.
Higgins, Jack (Fayette Co. fence mason and Guy family descendant). 1988. Interview with author, 29 July.
Higgs, Benny (Woodford Co. fence mason). 1989. Interview with author, 26 Jan.
Hilliard, Sam Bowers. 1984. *Atlas of antebellum Southern agriculture*. Baton Rouge: Louisiana State Univ. Press.
Hinkle, Susan T. (Bourbon Co. landowner). 1988. Letter to the authors, 11 July.
———. 1989. Letter to the authors, 16 Feb.
Hockensmith, Charles P. (Fayette Co. stonemason). 1989. Interview, Aug.
Hodgson, Simon F. (Development Director, British Trust for Conservation Volunteers). 1989. Letter to the authors, 27 June.
Hoffman, Charles G. 1835. *A winter in the West*. Vol. 2. New York: Harper.
Hofstra, Warren (Professor of History, Shenandoah College). 1988. Interview with author, May.
Holland, Robert. 1990. Well-wrought walls. *Audubon* (May): 48-57.
Holleran, Paul (Montgomery Co. descendant of fence masons William and Simon Holleran). 1990. Interview with author, 17 Sept.
Holmes, Hugh. 1820. On stone fences. *American Farmer* 1 (10 Mar.): 395-96.
Holt, James R. 1905. *One of the best Blue Grass farms in Kentucky*. Frankfort: privately printed.
Hopkins, Robert A. 1880. County Clerk's Office. Deed Book 65:19-20. Bourbon Co., Ky.
Houston, Cecil J., and William J. Smyth. 1990. *Irish emigration and Canadian settlement: Patterns, links, and letters*. Toronto: Univ. of Toronto Press.
Huff, James (Boyle Co. descendant of fence mason James Huff). 1989. Interview with author, Mar.
Huff, Mrs. Nick (Boyle Co. descendant of fence masons Jeff Carey and John McCrystal). 1989. Interview with author, 9 Mar.
Humphrey, H. N. 1916. *Cost of fencing farms in the north central states*. U.S. Department of Agriculture, Office of Farm Management, Bulletin no. 321. Washington: GPO.
Hunt, Michael O. (Indiana descendant of stonemasons Timothy Hunt and George Henry Hunt). 1990. Letter to the authors, 26 Sept.
Huskisson, Mike (Fayette Co. masonry contractor). 1988. Interview with author, 25 Oct.
Isaac, Rhys. 1982. *The transformation of Virginia, 1740-1790*. Chapel Hill: Univ. of North Carolina Press for the Institute of Early American History and Culture, Williamsburg, Virginia.
Isham, Doris (Fayette Co. descendant of fence mason Thomas P. Jones). 1989. Letter to the authors, 28 Aug.
Jenkins, John Geraint. 1965. *Traditional country craftsmen*. London: Praeger.
Jillson, Willard Rouse. [1926] 1972. *Old Kentucky entries and deeds*. Baltimore: Genealogical Publishing.
Jordan, Joe. 1940. *The Bluegrass horse country*. Lexington, Ky.: Transylvania Press.
Jordan, Terry G. 1966. *German seed in Texas soil: Immigrant farmers in nineteenth-century Texas*. Austin: Univ. of Texas Press.

Keightley, Frances (Mercer Co. historian). 1989. Interview with author, 3 Mar.

Kelley, M.A.R. 1940. *Farm fences*. U.S. Department of Agriculture, Farmer's Bulletin no. 1832. Washington: U.S. GPO.

Kelly, Stanley (Mercer Co. stonemason, restoration specialist). 1982. Interview with author, 25 Oct.

______. 1987. What yesterday left us. Unpublished manuscript, in possession of Stanley Kelly of Burgin, Ky.

______. 1989a. Interview with author, 26 Feb.

______. 1989b. Interview with author, 31 July.

______. 1989c. Interview with author, 24 Sept.

Kentucky Farmers Home Journal. 1941. Nov.

Kentucky Historical Society. Family sur-name files, unpublished manuscript collection. Frankfort.

King, William S. 1855. Letter to Brutus J. Clay, 17 Feb. Clay Family Papers. M.I. King Library, Special Collections, University of Kentucky, Lexington.

Klotter, James C. 1986. *The Breckinridges of Kentucky, 1760-1981*. Lexington: Univ. Press of Kentucky.

Kniffen, Fred, and Henry Glassie. 1966. Building in wood in the eastern United States: A time-place perspective. *Geographical Review* 56 (Jan.): 40-66.

Knight, Thomas A., and Nancy Lewis Greene. 1904. *Country estates of the Blue Grass*. Lexington, Ky.: Henry Clay.

Kramer, Carl E. 1986. *Capital on the Kentucky*. Frankfort: Historical Frankfort.

Landau, Mildred, and Joseph Landau. 1941. Old rock fences make new ones on U.S. 60. *Louisville Courier-Journal* 23 Feb.

Laws of Kentucky. 1799. Lexington: John Bradford.

Lee, Stephen (Woodford Co. fence mason). 1984. Interview with author, 15 Jan.

Leechman, Douglas. 1953. Good fences make good neighbours. *Canadian Geographical Journal* 47:218-35.

Lehman, William C. 1980. *Scottish and Scotch-Irish contributions to early American life and culture*. 2d ed. Washington: Lehman and Spahr.

Leible, Arthur B. 1952. Indiana, too: Limestone, transplanted Virginians—stone fences. *Louisville Courier-Journal Magazine* (4 May): 7.

Letton, George (Bourbon Co. resident). 1989. Interview with author, 10 Jan.

Lexington Herald-Leader. 1971. Cumberland Park helps restore skill of stonemasonry, a waning craft. 24 Oct.

______. 1982. Mending their fences. 11 May.

______. 1982. Stonewalling it. 20 July.

______. 1985. A water landmark. 13 Apr.

______. 1987. Longtome stone-fence builder Frank E. Guy, Sr., 77, dies. 17 July.

Long, Amos, Jr. 1961. Fences in rural Pennsylvania. *Pennsylvania Folklife* 12 (2): 33-35.

Louisville Courier-Journal. 1914. Sept. 13.

Lowenthal, David. 1968. The American scene. *Geographical Review* 58 (1): 61-88.

______. 1985. *The past is a foreign country*. Cambridge: Cambridge Univ. Press.

Lowenthal, David, and Hugh C. Prince. 1964. The English landscape. *Geographical Review* 54 (July): 309-45.

MacWeeney, Alen, and Richard Conniff. 1986. *Irish walls*. New York: Stewart, Tabori and Chang.

Maguire, John Francis. 1868. *The Irish in America*. New York: D. and J. Sadlier.

Manners, J.E. 1974. *Country crafts today.* Detroit: Gale.
Mannion, John J. 1974. *Irish settlements in eastern Canada: a study of cultural transfer and adaptation.* Toronto: Univ. of Toronto Press.
Martin, George A. [1887] 1974. *Fences, gates and bridges: A practical manual.* Brattleboro, Vt.: S. Greene.
Mastick, Patricia. 1976. Dry stone walling. *Indiana Folklore* 9:113-33.
Mather, Eugene Cotton, and John Fraser Hart. 1954. Fences and farms. *Geographical Review* 40:201-23.
McAboy [McAbay?], Mary R. T[?]. 1859. Letter to Mrs. Anne Clay. Clay Family Papers. M.I. King Library, Special Collections, Univ. of Kentucky, Lexington.
McClanahan, Bernice M. (Nicholas Co. resident). 1989. Interview, 12 Jan.
McDowell, Robert C., George G. Grabowski, Jr., and Samuel L. Moore. 1981. *Geologic Map of Kentucky.* United States Geological Survey and Kentucky Geological Survey.
McDowell, Samuel, ed. 1981. *Kentucky genealogy and biography.* Vol. 7. Louisville: Battey. Reprinted from J.H. Battle, W.H. Perrin, and G.G. Kniffen, *Kentucky, a history of the state,* 1888. Utica, N.Y.: McDowell.
McGrain, Preston. 1982. *Geology of construction materials in Kentucky.* Kentucky Geological Survey series 11, no. 13. Frankfort: Kentucky Geological Survey.
McGregor, Alfred E. (Fayette Co. building contractor). 1988. Interview with author, 25 Oct.
Mead, Andy. 1978. Much mystery, myth and romance surround old rock fences. *Lexington Herald and Leader* 12 Aug.
Mead, W.R. 1966. The study of field boundaries. *Geographische Zeitschrift* 54, no.2 (May): 101-17.
Melish, John. [1812] 1818. *Travels through the United States of America.* London: George Cowie.
Meredith, Mamie. 1951. The nomenclature of American pioneer fences. *Southern Folklore Quarterly* 15:109-51.
Metzger, Julie Schmitt (Nicholas Co. pastor). 1989. Letter to the authors, 13 Feb., quoting the Trustees Minute Book [Concord Christian Church] 4 Feb. 1854.
Miles, David (Bourbon Co. fence mason). 1989. Interview with author, 26 Apr.
Miller, J.R., Jr. (Fayette Co. civil engineer and descendant of stonemason Edmund P. Woods). 1989a. Letter to authors, 14 Feb.
———. 1989b. Letter to authors, 22 Apr.
———. 1989c. Interview with authors, 6 June.
Miller, Kerby A. 1985. *Emigrants and exiles: Ireland and the Irish exodus to North America.* New York: Oxford Univ. Press.
Miller, Miles (Bourbon Co. masonry contractor). 1989. Interview with authors, 25 Oct.
Mitchell, Arthur. 1880. *The past in the present.* Edinburgh: D. Douglas.
Mitchell, Brian. 1850. Bill of Sale, 7 Oct. Clay Family Papers. M.I. King Library, Special Collections, University of Kentucky, Lexington.
Mitchell, W.W. 1859. Bill, 3 Jan. Clay Family Papers. M.I. King Library, Special Collections, University of Kentucky, Lexington.
Mitchell, Donald G. 1875. Fences and division of farm lands. *Annual report of the secretary of the Connecticut Board of Agriculture,* 171-90. Hartford. Conn.: Case, Lockwood.
Mogge, Don (Bourbon Co. farmer). 1989. Interview with author, 2 Apr.

Moreland, K.M. 1989. Interview with author, 21 Feb.

Muir, Richard and Eric Duffey. 1984. *The Shell countryside book*. London: J.M. Dent.

Murray-Wooley, Carolyn. 1985. Kentucky River Marble as a building material. Paper presented at the Kentucky Images Conference, Kentucky Heritage Council. Louisville.

———. 1987. Early stone houses of Kentucky. Paper presented at the Historic Preservation Conference, Kentucky Heritage Council. Lexington.

———. 1988. Ulster-Irish in the Bluegrass. Paper presented at the Ulster-American Heritage Symposium, University of Ulster. Coleraine, Northern Ireland.

New York Times. 1914. 13 Sept.

Niles, Rena (Clark Co. landowner). 1984. Interview with author, 15 Jan.

Noble, Allen G. 1981. The diffusion of silos. *Landscape* 25, no. 1: 11-14.

Nourse, James. [1775] 1925. Journey to Kentucky in 1775. *Journal of American History* 19, no. 2: 252.

O Borchgrevink, A.B. 1982. A Dumfriesshire drystone waller. In *Gold under the furze,* edited by R.A. Gailey and D. O'Hogain. Dublin: Glendale.

O'Connor, Clem J. 1952. Monuments to leisure. *Louisville Courier-Journal Magazine* 4 May, 5-7.

O Danachair, Caoimhin. 1957. Materials and methods in Irish traditional building. *Journal of the Royal Society of Antiquaries of Ireland* 87:61-74.

O'Neill, Timothy. 1984. Stonecutting and masonry. In *Ireland's Traditional Crafts,* edited by David Shaw-Smith. London: Thames and Hudson.

Parrish, D. [Dabney] W. 1836-1882. Daybook. D.W. Parrish Papers. M.I. King Library, Special Collections, University of Kentucky, Lexington.

Patmore, J.A. 1974. Routeways and recreation patterns. In *Recreational Geography,* edited by Patrick Lavory. New York: John Wiley.

Peet, Henry J., ed. 1883. *Chaumiere papers; Containing matters of interest to the descendants of David Meade. . . .* Chicago: Horace O'Donoghue.

Perrin, William Henry. 1882a. *History of Bourbon, Scott, Harrison, and Nicholas Counties, Kentucky.* Chicago: O.L. Baskin.

Perrin, William Henry. 1882b. *History of Fayette County, Kentucky.* Chicago: O.L. Baskin.

Perrin, William Henry, et al. 1888. *Kentucky: History of the state.* Louisville: Battey.

Pinkett, Harold T. 1977. Leadership in American agriculture: The published documentary heritage. In *Agricultural literature: Proud heritage—future promise,* edited by Alan Fusonie and Leila Moran, 159-68. Washington: Association of the National Agricultural Library and United States Department of Agriculture.

Placilla, Libby. 1988. Law is needed to protect rock fences. *Lexington Herald-Leader,* 24 Feb., A-10.

Pocius, Gerald L. 1977. Walls and fences in Susquehanna County, Pennsylvania. *Pennsylvania Folklife* 26:9-20.

Prunty, Merle, Jr. 1955. The renaissance of the Southern plantation. *Geographical Review* 45 (4): 459-91.

R.R. 1956. Stone fences are at once a boon and a problem to farm owners. *Lexington Herald-Leader,* 15 Jan.

Rackham, Oliver. 1986. *The history of the countryside*. London: J.M. Dent.

Railey, William E. [1938] 1968. *History of Woodford County.* Lexington, Ky.: Thoroughbred.

Rainsford-Hannay, Frederick. 1957. *Dry Stone Walling*. London: Faber and Faber.
Raistrick, Arthur. [1946] 1988. *Pennine walls*. Lancaster, England: Dalesman.
Raup, H.F. 1947. The fence in the cultural landscape. *Western Folklore* 6:7.
Rebmann, James (Fayette Co. city planner). 1988. Interview with author, 15 Oct.
Rhodes, Chip. 1989. History written in stone. *Kentucky Advocate* (Danville) 9 Apr.
Richardson, Alfred J., Rudy Forsythe, and Hubert B. Odor. 1982. *Soil survey of Bourbon and Nicholas Counties, Kentucky*. Washington: U.S. Department of Agriculture Soil Conservation Service.
Richardson, Charles Henry. 1923. *The building stones of Kentucky*. Kentucky Geological Survey, series 6, vol. 11. Frankfort: Kentucky Geological Survey.
Rice, Otis K. 1951. Importation of cattle into Kentucky, 1785-1860. *Register of the Kentucky Historical Society* 49:35.
Rion and Mitchell. 1848. Bill, 23 Oct. Clay Family Papers. M.I. King Library, Special Collections, University of Kentucky, Lexington.
Rion and Sharrard. 1851. Bill, 31 March. Clay Family Papers. M.I. King Library, Special Collections, University of Kentucky, Lexington.
Roberts, Isaac Phillips. 1905. *The farmstead: The making of the rural home and the lay-out of the farm*. New York: Macmillan.
Robinson, Philip. 1984. *The plantation of Ulster: British settlement in an Irish landscape, 1600-1670*. Dublin: Gill and MacMillan.
———. 1989. Letter to the authors, 12 Jan.
———. 1990. Letter to the authors, 5 Jan.
Rollinson, William. 1967. A history of man in the Lake District. London: Dent.
———. [1969] 1972. *Lakeland walls*. Yorkshire, England: Dalesman.
Rusticus [Cassius M. Clay]. 1859. Water gaps. *Kentucky Farmer* 1 Jan.:122-23.
San Francisco Examiner. 1914. 13 Sept.
Schwaab, Eugene L., ed. 1973. *Travels in the old South*. Vol. 7. Lexington: Univ. Press of Kentucky.
Shaker Ledger Books. 1839-1871. Filson Club Library, Louisville. Microfilm reel 5, vols. 25, 26, read by Christopher Williams and Bill Kephart.
Shaughnessy, Andrew. 1986. A Yorkshire legacy. *Blair and Ketchum's country journal* 13 (May): 34-43.
Simkins, Jacqui (Secretary, Dry Stone Walling Association of Great Britain). 1989a. Letter to authors, 3 June.
———. 1989b. Letter to authors, 7 July.
———. 1989c. Letter to authors, 16 Nov.
Smith, Marvin (Washington Co. fence mason). 1989. Letter to authors, 15 Feb.
Smith, William Bradley (Mercer Co. landowner). 1989. Interview with author, 15 May.
Smith, William W. (Bourbon Co. farm owner). 1989. Interview with authors, Jan. *Soil of the South*. 1854. Fences. 4:6.
Soper, Johnny (Bourbon Co. farm owner). 1988. Interview with author, 24 Oct.
Soper, Tom (Bourbon Co. farmer and stonemason). 1990. Interview with authors, 11 Mar.
———. 1988. Interview with author, 2 Nov.
Southern Planter. 1858. Stone fencing. Reprinted in *New England Farmer* 10 (Nov.): 496-97.
Spengler, Frances Thornton (Fayette Co. Thornton family descendant). 1990. Interview with authors, 13 Feb.

Stallons, Malcolm. 1976. Laying stone fences like a jigsaw puzzle. *Lexington Herald-Leader* 26 Mar.
Standiford, Thelma (Nicholas Co. resident). 1989. Interview with author, Jan.
Stephens, Henry. 1877. *The book of the farm*. Vol. 2. London: W. Blackwood.
Straight, Stephen. 1987. Stone walls. *Pioneer America Society Transactions* 10:67-75.
Swasy, Alecia. 1989. Bluegrass country of Kentucky goes to shopping malls. *Wall Street Journal* 5 Apr., 1.
Tate, Sarah House (Fayette Co. architect). 1984. Letter to the author, 21 Feb.
Taylor, Ed (Woodford Co. fence mason). 1988. Interview with author, Oct.
Taylor, Richard (Franklin Co. farm owner). 1988. Interview with author, 13 Jan.
Thierman, Sue McClelland. 1959. Good Guys. *Louisville Courier-Journal Magazine* 22 Nov.:44.
Thomas, Samuel W., and James C. Thomas. 1973. *The simple spirit*. Shakertown at Pleasant Hill, Ky.: Pleasant Hill Press.
Thoroughbred Record. 1926. 104, no.21: 318.
Thwaites, Reuben Gold, ed. 1904. *Early western travels, 1784-1846*. Vol. 10. Cleveland: Arthur H. Clark.
Tischendorf, Alfred, and E. Taylor Parks. 1964. *Diary and journal of Richard Clough Anderson*. Durham, N.C.: Duke Univ. Press.
Traynor, Retha (Mercer Co. great-niece of stonemason Mike McCrystal). 1989. Interview with author, 9 Mar.
Troutman, Richard L. 1968. The physical setting of the Bluegrass planter. *Register of the Kentucky Historical Society* 66 (Oct.): 367-77.
Tufnell, Richard (Scotland, Certified Master Craftsman). 1989. Letter to the authors, 5 June.
———. 1990. Letter to the authors, 16 Sept.
Tuttle, Lena (Fayette Co. descendant of fence mason William Dargavell). 1991. Interview with author, 24 Oct.
U.S. Census Bureau. 1850-1880. Population Censuses, Bourbon and Woodford counties, Ky. Manuscripts on microfilm. Washington.
U.S. Census Bureau. 1860. Agricultural Census, Eastern Division, Bourbon Co., Ky. Manuscripts on microfilm. Washington.
U.S. Commissioner of Agriculture. 1871. Statistics of fences in the United States. *Report of the Commissioner* 497-512. Washington: U.S. Dept. of Agriculture.
Valley Farmer. 1857. Farm fences—stone walls. 9, no. 42: 42-43.
Vanderhoef, Jewel. 1989. Letter to the editor. *Lexington Herald-Leader* 17 June.
Venable, John (Clark Co. landowner). 1988. Interview with authors, 24 Oct.
———. 1989. Interview with authors, 20 Feb.
Vigne, Godfrey T. 1832. *Six months in America*. Vol. 2. London: Whittaker, Treacher.
Vince, John. 1983. *Old farms: An illustrated guide*. New York: Schocken.
Vivian, John. 1976. *Building stone walls*. Charlotte, Vt.: Garden Way.
Warder, John A. 1858. *Hedges and evergreens*. New York: Orange Judd.
Waugh, Bobby (Scott Co. fence mason). 1988. Interview with author, 25 Oct.
———. 1989. Letter to authors, 11 Oct.
Waugh, Ernest, Jr. (Bourbon Co. mason). 1989. Interview with authors, Jan.
Waugh, Larry (Bourbon Co. fence mason). 1989. Interview with author, 15 Jan.
Webster's new international dictionary of the English language. 1923. Springfield, Mass.: Merriam.

Wells, Ron (Executive Director, Woodford Co. Historical Society). 1988. Interview with authors, Apr.

Whitley, Edna Talbott. 1956. *Kentucky ante-bellum portraiture.* N.p. National Society of Colonial Dames in America.

Wickliffe, C.A., S. Turner, and S.S. Nicholas. 1852. *Revised statutes of Kentucky.* Frankfort: A.G. Hodges, state printer.

Widener, P.A.B. 1940. *Without drums.* New York: Putnam.

Wiles, Howard (Scott Co. farm owner). 1988. Interview with authors, 19 Oct.

Wilgus and Bruce Company. 1855. Bill, 25 Jan. Clay Family Papers. M.I. King Library, Special Collections, University of Kentucky, Lexington.

Williamson, Duane Edward. 1967. A study of the stone fence as a landscape feature of Fayette County, Kentucky. Master's thesis, University of Kentucky.

Willmott, George F., Jr. (Fayette Co. landowner). 1989. Interview with author.

Wilson, Mary Allan (Robertson Co. landowner). 1989. Letter to the authors, 15 June.

Wilson, Samuel M. 1933. Pioneer Kentucky in its ethnological aspect. *Register of the Kentucky Historical Society* 31 (Oct.): 283-95.

Witt, Malcolm (Bourbon Co. masonry contractor). 1990. Interview with author, 18 Oct.

Wolfe, Robert (Clark Co. stonemason). 1990. Interview with author, 30 Sept.

Woods, John. [1821] 1904. Two years' residence in the settlement on the English prairie, in the Illinois Country, United States, 5 June 1820 to 3 July 1821. In *Early western travels, 1748-1846.* Vol. 10, edited by Reuben Gold Thwaites. Cleveland: Arthur H. Clark.

Wooley, R. Matthew (Fayette Co. carpentry contractor). 1990. Interview with author, 2 Oct.

Worford, Charles (Anderson Co. fence mason and Worford family descendant). 1988. Interview with authors, 7 Sept.

Young, Bennett H. 1898. *History of Jessamine County, Kentucky.* Louisville: Courier-Journal Job Printing.

Zelinsky, Wilbur. 1959. Walls and fences. *Landscape* 8:14-20.

Index

The text includes figures and tables. Locators that include an F indicate a figure, and locators that include a T indicate a table (e.g., 45T indicates a table on page 45). In some cases, the same endnote number, but in subsequent chapters, appears on a single page. In locators for these endnotes, the chapter number precedes the note number (e.g., 199 n 2.1 indicates note 1 in chapter 2, which is located on page 199).